Photoresist

Photoresist

MATERIALS AND PROCESSES

W. S. DeForest

Autonetics Division, Rockwell International

McGRAW-HILL BOOK COMPANY

New York St. Louis San Francisco Auckland Düsseldorf Johannesburg Kuala Lumpur London Mexico Montreal New Delhi Panama Paris São Paulo Singapore Sydney Tokyo Toronto

Library of Congress Cataloging in Publication Data

DeForest, William S
Photoresist : materials and processes.

1. Photoresist. 2. Integrated circuits.
3. Printed circuits. I. Title.
TK7874.D42 621.381'73 74-20636
ISBN 0-07-016230-1

1234567890 KPKP 784321098765

The editors for this book were Jeremy Robinson and Margaret Lamb, the designer was Naomi Auerbach, and the production supervisor was George Oechsner.

Printed and bound by The Kingsport Press.

To my wife, Dede, for her patience.

Contents

Preface

Photoresist has grown in importance along with the electronics industry, since even the simplest electronic systems utilize photoresist in their manufacture. Engineers and production personnel must deal with many problems that arise daily with photoresist, as with any proprietary product. This book was written primarily to help solve those problems, as well as to assist in setting up developmental processes and for more efficient operation.

The initial chapter discusses early photoresists and how they were used in the printing industry and follows the early transfer of technology to the electronics industry. Basic circuit-board and multilayer-board technology is touched on and illustrated.

In Chapter 2, the means by which different types of resists function is considered. Resist constituents and how they work are discussed based upon available information. Although this chapter on practical theory is not essential to the use of the book, it is especially worthwhile for those interested in microelectronics.

Heavy emphasis is placed upon metal cleaning and surface preparation, a critical but sometimes neglected step. Chapter 3 is devoted to metal cleaning and preparation for photoresist application. Examples of cleaning cycles for various metals are provided, with the emphasis on copper.

The entire processing sequence is thoroughly spanned, from application methods, exposure, and developing to touchup and removal. In Chapter 4, practical processing information is given for negative-acting wet-film resists such as Kodak Photoresist (KPR). Chapter 5 deals with positive resists such as the Shipley AZ-119. Emphasis is placed upon roller coating. Chapter 6 is devoted to dry-film resists and includes both compositional and processing information. Microelectronics, which has unique needs, is covered in Chapter 7, which contains information for positive and negative resist processing, including cleaning, for microelectronic applications.

Cross referencing has been held to a minimum to make each chapter "stand alone" as much as practical. The book also contains a great deal of reference material for those so inclined and can be used as a practical text in photoresist matters.

No chemical analysis of any proprietary product has been made in the preparation of the book; all opinions are the author's and are based upon interpretation of the available information.

W. S. DeForest

Acknowledgments

It would be impossible to acknowledge all those who have assisted in the preparation of this book. I would, however, like to mention those who were instrumental in this publication. First, my employer, Rockwell International, deserves special thanks for granting permission to publish the book; J. J. Licari, B. F. Barnett, H. V. Connelly, and L. L. Peart, all of Rockwell International, provided valuable assistance and encouragement. The following persons from the photoresist industry were also instrumental in this publication:

Bob Burr of
Bell and Howell Corp.

Jim Barton and
Chris Johnson of the
Riston Products Division
of Du Pont

Mel Libson,
Gene Wiener,
and Vern Kurisu of
Dynachem Corporation

Mark Mahon of
Gyrex Corporation

Jack Murphy and
Bob Stens of
P. A. Hunt Chemical Company

Mike Gulla of
Shipley Company

Joel Davis
Larry Burgess
Joan Couger

Photoresist

CHAPTER ONE

History of Photoresists

Present-day photoresists and photoetching processes evolved largely from technology developed in the printing industry. Circuit boards and microcircuits were undreamed of in 1852 when W. H. F. Talbot patented a photoetching process that could be used on copper (British Patent 565). The photoresist used was gelatin sensitized with a bichromate (or dichromate) salt; ferric chloride solution served as the etchant. Even earlier, in 1826, J. N. Niepce had found that certain types of asphalt were photosensitive, and he had successfully etched patterns in pewter. This discovery, however, never reached the importance of Talbot's.

Perhaps the first of the natural organic colloids to be insolubilized was starch (1) (A. E. Becquerel, 1840), but subsequently proteins, carbohydrates, and shellac were found to be capable of sensitization by dichromates. To be useful as a photoresist, the material had to be capable of light sensitization by dichromates, had to have chemical resistance and film-forming properties, had to be reasonably adherent to metal surfaces, and preferably had to be water-soluble in the unexposed or unhardened condition for ease of application and developing. Proteins, such as animal and fish glues, blood and egg albumen, casein, and gelatin, met most of these criteria.

Carbohydrates, such as polysaccharide gums (gum arabic), and starches were also found useful. The coating solutions were prepared by dissolving the polymer and ammonium dichromate in water. Additives to control the pH, retard spoilage, and to achieve various other properties were sometimes used. Potassium dichromate proved to be an adequate substitute for the ammonium salt, but the sodium salt was generally avoided because of its deliquescent nature. Chromate salts, as distinguished from dichromates, are reportedly (2) of limited value.

Dichromate-sensitized resists are generally negative-acting; areas which are exposed to light in the ultraviolet range are rendered insoluble in water, whereas the unexposed areas still retain their solubility. Exposed areas undergo a color change because of the reduction of some of the dichromate salts, and the light-hardened areas are sometimes referred to as being "tanned." After exposure, the parts are developed, or "stripped," to form the image. The process is illustrated in Fig. 1-1.

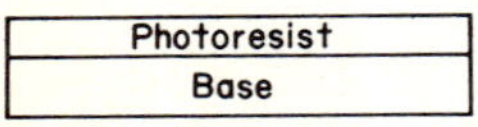	Negative-acting photoresist is applied uniformly in a thin layer on the circuit board or other base material and dried thoroughly.
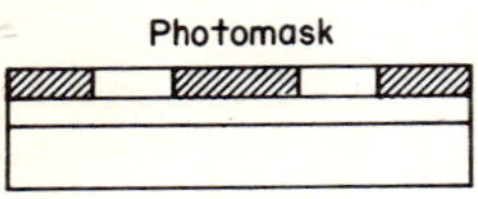	A photomask with transparent image areas delineating the desired pattern is brought in close contact with the photoresist surface through the use of a vacuum frame or by mechanical means.
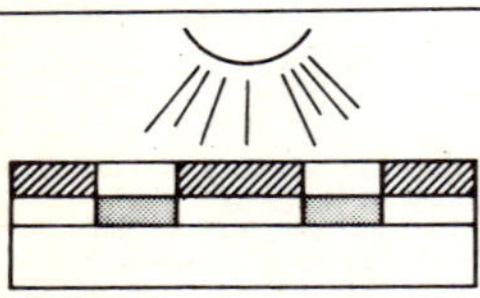	With the photomask in place, the resist is exposed to a light source rich in ultraviolet radiation. The resist film beneath the clear areas of the photomask undergoes a physical and chemical change that renders it insoluble in the developing solution.
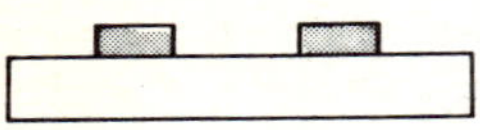	After exposure, the resist-coated base is immersed in the developing solution, which allows the unexposed areas to be removed without excessive effect on the hardened or exposed area. The resulting image can be further hardened by baking or rendered more visible by dyeing or both. The base can be etched or plated (Fig. 1-5).

Fig. 1-1 Negative-acting photoresist process.

Positive resists are relatively new, though one of the primary components, the photosensitizers, was used in other forms of reproduction for many years. Aromatic orthoquinone diazides, for example, were prepared by P. J. Griess in 1858 (3), almost 100 years before the introduction of positive-acting photoresists containing them. Diazo prints on paper were made in the 1890s, and in 1899 A. and L. Lumière (4) noted that positive images could be made with diazo compounds and organic colloids, such as casein. In 1923, Kalle and Co. Aktiengesellschaft introduced the Ozalid reproduction process for paper; the process utilized photosensitive diazo compounds, which formed colored dyes in unexposed areas by chemically joining or coupling with various other compounds called "coupling agents" during developing with moist ammonia. Areas exposed to light could not undergo the coupling reaction because the diazo compound had been decomposed and deactivated by the light. This principle is still used today in positive-working photoresists except that the coupler is a resin that is soluble in the developer in areas where it has not reacted with the diazo sensitizer. Couplers are usually phenol derivatives. A typical positive process is shown in Fig. 1-2.

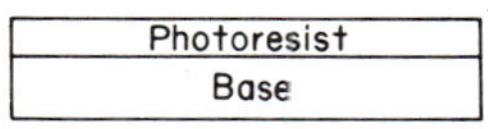

Positive-acting photoresist is applied uniformly in a thin layer on the circuit board or other base material and dried thoroughly.

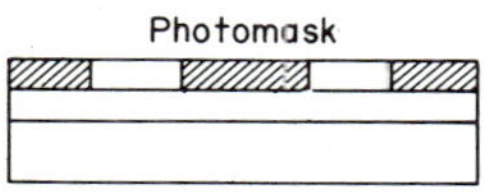

A photomask with opaque image areas delineating the desired pattern is brought in close contact with the photoresist surface through the use of a vacuum frame or by mechanical means.

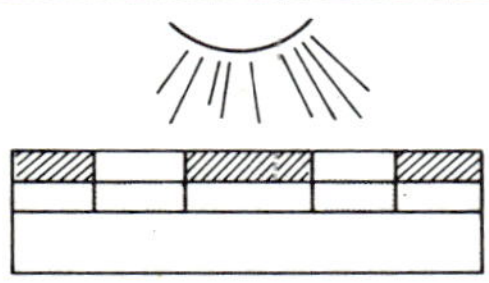

With the photomask in place, the resist is exposed to a light source rich in ultraviolet radiation. The resist film beneath the clear areas of the photomask undergoes a chemical change that renders it soluble in the developing solution.

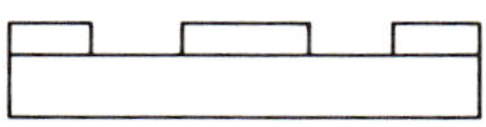

After exposure, the resist-coated base is immersed in the developing solution, which dissolves the exposed areas. The resist images may then be baked for greater chemical resistance. Positive resist images are normally not dyed.

Fig. 1-2 Positive-acting photoresist process.

Most positive photoresists originated from the early lithographic plate work at Kalle Co. AG in Germany and its American subsidiary, Azoplate Corporation of Murry Hill, New Jersey.

The first of what will be termed "modern-day" photoresists, KPR (Kodak photoresist), utilizes cinnamic acid derivatives, which are intrinsically photosensitive compounds. In 1895, J. Berthram and K. Kursten first observed the sensitivity of the acid, and later investigators established that the cross linking occurred between the unsaturated ethylenic linkages. However, free cinnamic acid was not capable of forming continuous films, a primary photoresist requirement. In 1931, Alexander Murry of Kodak (5) noted that when mixed with natural film-forming gums and resins, cinnamic acid derivatives were hardened on exposure to an ultraviolet light source, and a useful image could be formed. From observations, it was presumed that the resins took no part in the reactions but acted to prevent the cinnamate compound from crystallizing. In their noncrystalline state, the cinnamates were able to polymerize when exposed to light in the ultraviolet range. The coating solution was prepared by dissolving the cinnamate derivatives and resins in acetone and benzol. Basic resins mentioned were copal, cumar, pontianak copal, kauri gum, and others with small amounts of cellulose nitrate added to reduce tackiness. Plates suitable for etching were prepared by exposing coated surfaces, developing with mineral oil and cotton swabs, followed by a soap-and-water wash. Copper images could be etched with ferric chloride or nitric acid using the exposed image as a resist. Lithographic plates could also be prepared with the mixture because of its ink-receptive nature.

Work on cinnamates continued through the 1940s and 1950s. Cinnamate compounds, usually cinnamoyl chloride, could be reacted with polymeric materials to form light-sensitive esters and ketone (see Chap. 2) with high molecular weight, good solvency, and excellent film-forming properties. Some examples are shown in Fig. 1-3. The KPR resin is the

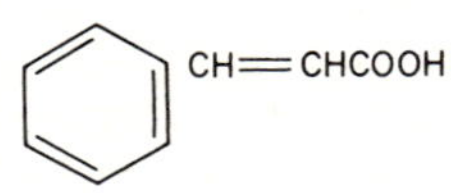

Cinnamic acid

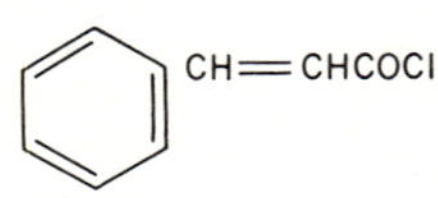

Cinnamyl chloride

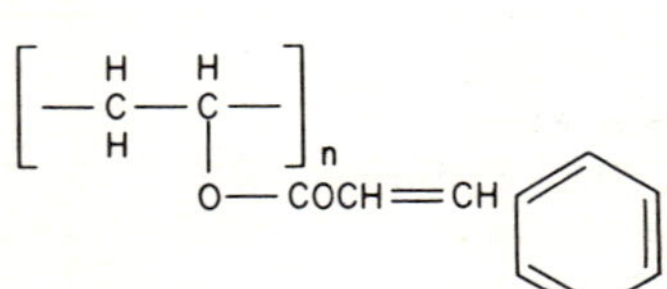

Polyvinyl cinnamate

Fig. 1-3 Polyvinyl cinnamate and related compounds.

cinnamate ester of polyvinyl alcohol (7–10). Its introduction by Kodak in about 1953 marked the beginning of an era which would see the evolution of over 40 photoresists designed to serve the rapidly expanding electronics industry within the next 17 or so years. Table 1-1 lists those resists known to the author in early 1972, along with their respective manufacturers or suppliers. It may be safely stated that these resists are a cornerstone in the fabrication processes of nearly all electronic equipment manufactured in this country and that the phenomenal growth of the electronics industry of the last two decades is due in no small part to the resist's unique ability to accurately reproduce chemically resistant images on metal and oxide surfaces.

TABLE 1-1 Photoresists Offered as of 1971–1972

Kodak[a]	Shipley[b]	Dynachem[c]	Du Pont[d]
KPL	AZ-15	DCR #3140	Riston 30S
KPR	AZ-17	DCR #3116	Riston 16S
KPR-2	AZ-340	DCR #3118	Riston 211
KPR-3	AZ-345	DCR #3118-H	Riston 14
KPR-4	AZ-1350	DCR #3154	Riston 25
KMER	AZ-111	DCR #3170	Riston 6
KTFR	AZ-119	DCR LS	Riston 15S
KOR	AZ-1350H (J)	DCR Laminar Series	Riston 218
....	....	DCR Laminar AP	Riston III Series
KAR-3	....	DCR Laminar A	Riston II Series
....	....	PRS-175	Riston 50
747	....	CMR Series	
....	....	Micro-G	....
GAF[e]	**Meaker[f]**	**Hunt[g]**	**Brady[h]**
Microline PR-102	HSR-160	Waycoat PF	Phodar 10
Microline PR-115	HSR-200	Waycoat 20	Phodar 15
Microline PR-207	....	Waycoat RC	Phodar 25
Microline PR-301	....	Waycoat IC	....
....	....	Waycoat HR200	....
....	....	Waycoat SC	....
....	....	Waycoat 450	....

[a] Eastman Kodak Company, Rochester, N.Y.
[b] Shipley Company, Newton, Mass.
[c] Dynachem Corporation, Santa Fe Springs, Calif.
[d] E. I. du Pont de Nemours & Company, Wilmington, Del.
[e] GAF Corporation, New York, N.Y.
[f] The Meaker Company, Nutley, N.J. (discontinued).
[g] P. A. Hunt Company, Palisades Park, N.J.
[h] W. H. Brady Company (Photopolymer Research Corp.), Milwaukee, Wis. (discontinued).

CIRCUIT BOARDS AND PHOTORESISTS

The circuit-board and photoresist technologies have grown together in stature and importance. Both are an outgrowth of the printing industry, where photoresists have been known for more than a century. In early circuit-board processes, etch resists were applied by offset lithographic printing, and the natural transfer of technology followed. To better understand the photoresist developments, a brief discussion of the circuit-board and electronics industries for which they were developed will be given.

Circuit-Board Processes

In 1925, Ducas patented a method (11) of forming electroplated conductor patterns in copper, gold, or silver on a nonconductive base material. The conductor patterns were formed by stenciling rather than photoetching, but the patterns are of interest because of their remarkable resemblance to modern-day circuit patterns (Fig. 1-4). A conductive paste was stenciled on a plastic base and electroplated with the aforementioned metals.

The conductors were designed to interconnect circuit elements in lieu of wires, while the base acted as an insulator. To quote a claim in part, a method was described for "constructing electrical circuits arranged ... upon the panel surfaces ... for interconnecting terminals of electrical instruments secured to [the] surfaces."[1] There is no evidence that extensive commercial use was made of this invention, or of any subsequent circuit-board invention, until World War II.

In England in 1936, Eisler (12) conceived what is probably the earliest printed circuit boards as we know them. In the early 1940s, despite the need for electronic equipment and greater production quotas brought about by the war, little interest was shown by equipment manufacturers. On February 3, 1944, he filed a patent in the United States describing a method of manufacturing a component of an electric circuit. The manufacturing process involved printing the circuit pattern on a foil bonded to an impervious insulating backing using a medium (ink) which was adherent to the foil and capable of protecting the covered metal surfaces from chemical attack. The pattern was formed where the unprotected foil was etched away to the base. Today, this process is generally referred to as "print and etch" (Fig. 1-5). In 1948, the patent was granted (13), some 4 years after filing. Eisler continued his work in the area of circuit boards (14–16) and filed other patents.

In 1947, it was revealed that larger numbers of circuit boards had been

[1] C. Ducas, U.S. Patent 1,563,731 (1925).

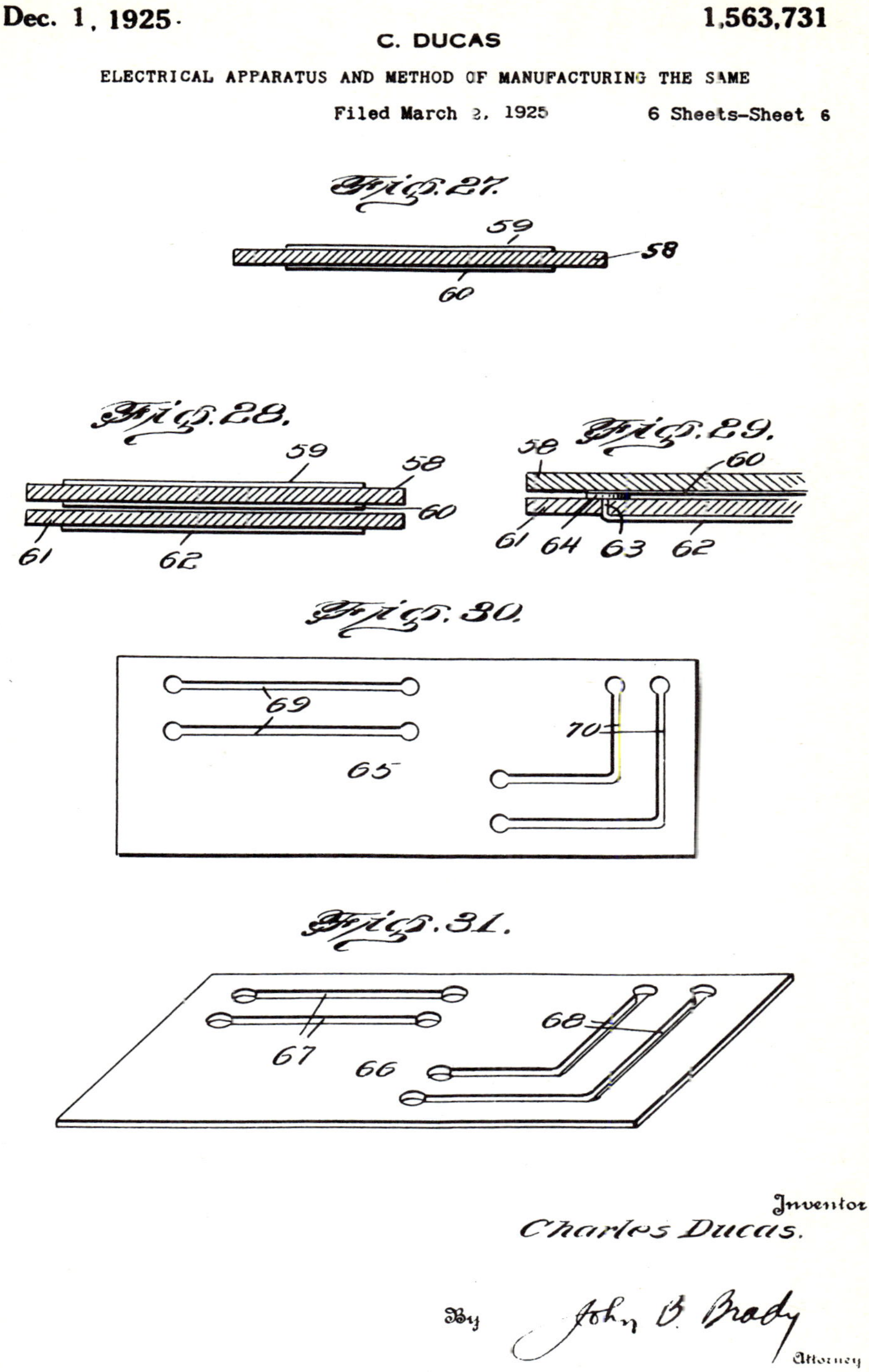

Fig. 1-4 Early patent (1925) on forming circuit-board conductors.

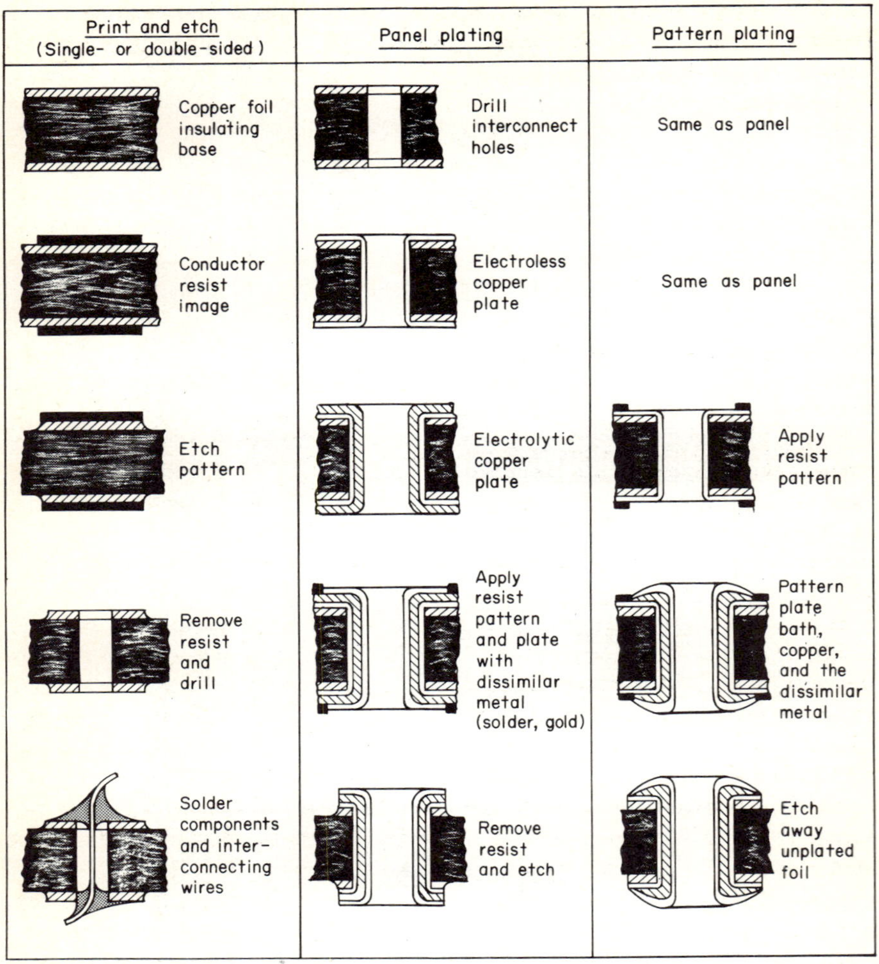

Fig. 1-5 Basic circuit-board processes. (Also see Fig. 6-2).

manufactured for use in ordnance devices in the United States. Interest grew rapidly and numerous processes were evolved by the early 1950s. Mechanical fabrication techniques (17–24) were quite popular for forming both circuitry and interconnections during this early period, giving way to plated-foil processes in the latter part of the decade (25). The use of eyelets (17,22,24) for interconnnections was the accepted method until the advent of suitable electroless processes for metallizing the through holes. Conductive ink patterns (26–30) applied by stenciling or screening also found some acceptance and were used in both the plated and unplated conditions until the late 1950s. Although adequate for ceramic substrates (what is now termed "thick films"), the early inks with resin binders and conductive metal particles were not very successful (31) on plastic bases.

Plating as a means of interconnecting both sides of a circuit pattern evolved somewhat slowly during the decade in spite of the fact that methods for metallizing the through holes were available. Wein (32) in his survey of copper-metallizing processes for nonconductors found that the two-step catalyzing process for electroless copper deposition was available in 1948. Glass or other nonconductive surfaces were sensitized with a dilute stannous chloride solution which acted as a mordant and to reduce a precious metal such as gold from a subsequent acid-activating solution. Wein credited Dr. Henry S. Myers with the process. Narcus in patents filed in 1946 had described the use of stannous compounds as a sensitizer (33) for silvering nonconductive and a copper-metallizing process (34) which utilizes a titanium compound as a mordant and palladium chloride as the activation for an electroless copper solution. Metallization, in general, of plastic nonconductors had long been known (35,36). The primary reason for the delay was the instability, cost, and unreliability of the early electroless copper solutions. Decomposition in a few hours or less was hardly a recommendation for high-volume production. Other methods for forming a conductive path from one side of the board to the other were tried (37,38,18,22) as was electroless nickel (39,40), but apparently with limited success [though low-temperature electroless nickel baths (41) can be used in some processes]. With the questionable reliability of eyelets (42–44) looming as an incentive, considerable effort was expended toward developing a reliable electroless copper suitable for production use. Atkinson (45) in 1956 filed a patent for a very successful activating and plating process. Other processes followed (46–52), including the electroless copper used at Autonetics (53) for many years. In 1960, the Shipley Company introduced the colloidal catalyst (52,54), which quickly became the standard sensitizing-activating process for the printed-circuit industry; it remains so today.

In the 1960s, manufacturing processes for printed circuit boards stabilized into three basic processes—the original print-and-etch method, the panel-plating method, and the pattern-plating method, as illustrated in Fig. 1-5.

In the early and mid-1960s, multilayer boards began to increase in importance, particularly for military uses, where the cost was secondary to compactness, weight, and reliability. Multilayer boards evolved with advances which occurred in the metallizing and photoresist processes. Like the conventional boards, both mechanical and plated interconnection methods were tried, with the latter the surviving process. Some examples of various types are given (55,56,29), but depending on the design there are only two widely used processes for manufacturing multilayer boards.

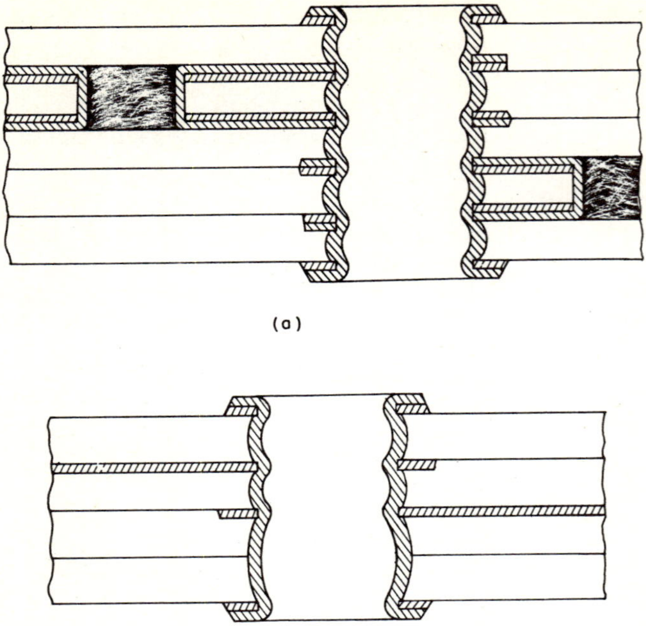

(a)

(b)

Fig. 1-6 Multilayer processes. (*a*) Multilayer board design with secondary interconnections; (*b*) multilayer board design with through-hole interconnection.

In the simpler process (Fig. 1-6), all interconnections go completely through all layers of the board, whereas the more complicated boards have discrete connections offset from the through-hole connections.

Standards (57–59) were prepared for conventional boards in the early 1960s and in the latter part of the decade for multilayers (60). Silver was permanently eliminated as competition for copper in metallizing circuit-board holes because of the migration problems (61,62) and was restricted in MIL-STD-275A (paragraph 5.9.1) for military boards (58). Various innovations from the conventional methods have been proposed. Harmon (63) discussed and compared unique methods for manufacturing multilayer boards. Nagy (64) patented a method for making padless circuitry, while Nierenberg (65) patented methods for fabricating boards with flushed circuitry. All-copper circuitry (no dissimilar metals) has been made possible by special photoresists. In the Shipley CTH process (66), a positive resist replaces the dissimilar metal in the plated through holes as an etch resist. The resist, Shipley AZ-345 or AZ-340, is especially formulated for this purpose. Du Pont Riston and Dynachem Laminar can be used to seal off holes and pad areas while etching. The process is termed "tenting" (67).

Resist Processes

The early selection of offset lithographic printing as a means of applying etching resists was quite natural when one considers the nature of the inks that must be used for lithography and the other available methods. Lithography is one of the three basic printing methods, the other two being intaglio (gravure, or engraving) and typographic processes. Figure 1-7 shows the basic differences in the three printing methods. Photoetching of copper with ferric chloride is a natural evolution of the engraving art and

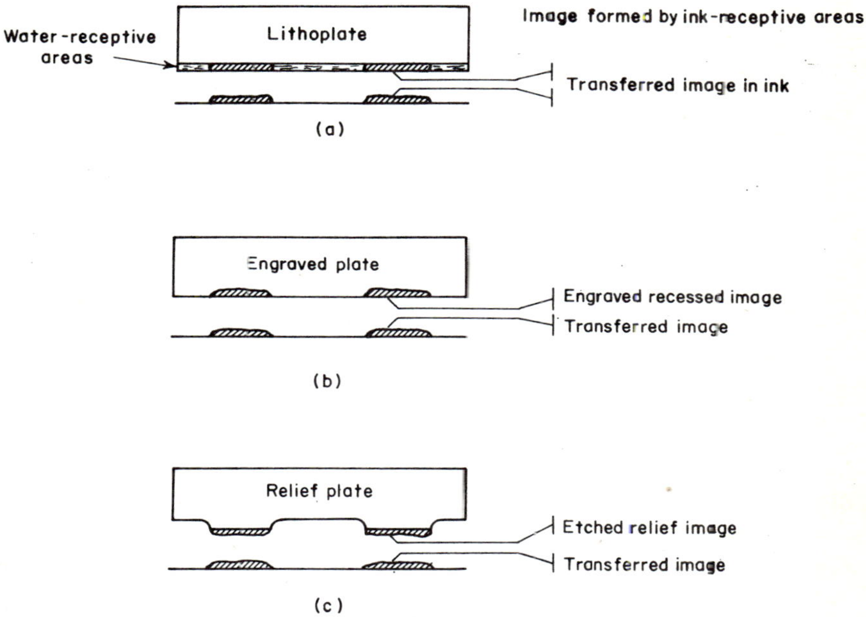

Fig. 1-7 Basic printing processes. (*a*) Lithographic, or planographic, method; (*b*) intaglio method (includes engraving and gravure); (*c*) typographic, or relief, method.

has been known for over 100 years. Today, it is still used with few basic changes except for the nature of the resist. Lithography depends upon the interfacial tension of oily inks and water-retaining surfaces for image formation. The inks, then, are naturally hydrophobic and can be made resistant to the action of etchants when dry. Offset lithography refers to a printing method in which the inked lithographic plate prints the image on a rubber blanket or roller, which in turn prints the image on the object (for example, a circuit board). The rubber image is better able to conform to uneven surfaces than the metal lithographic plate, and very little moisture is transferred with the ink compared with direct lithoprinting. Copper is quite receptive to most printing inks and is often used in lithographic plates.

Plating and etching resists are still being applied to circuit boards by offset methods with some success (68). Information markings on circuit boards to designate component mounting positions and board part numbers have been printed (69). Applying resist images by offset printing has, at present, several problems that must be overcome: (1) The process results in thin coatings (80 μin. or less), which usually necessitates two or more passes; (2) coatings are generally nonuniform, particularly when used for masking large areas; (3) in some instances misalignment and poor registration occur.

Improved ink (resist) characteristics can reduce these problems to some extent, and future formulations may allow increased use of this rapid, inexpensive reproduction method. At present, it can be used to produce boards of moderate quality once the art is mastered. In his report, Woods (68) indicated that the printed resist could be fortified after the image had been formed by applying powdered fusible resins, such as bitumen or dragon's blood, to the tacky resist surface and baking at a temperature sufficiently high to cause curing and fusing. Acceptable chemical resistance is acquired in this manner, though the time consumed by powdering and baking largely offsets the gains accrued by the printing methods. The most feasible approach would appear to be through improved printable resists. Some efforts have been expended in this area and improvements have been made. However, it is unlikely that this method will replace screening methods or photoresists in the near future.

Several other methods have been used to apply resists to circuit boards. The use of decalcomania was proposed by Gillett and Katz (70) and the Haloid Company (71) as an etch resist. Haloid also tested xerography. Pattern formation by forcing ink or resist through a stencil (Fig. 1-5) was used by Ducas (11), Cardy (26), and others (28).

Resist application by screening was a success from the very beginning of circuit-board manufacturing. Other than photoresists, screening is the only method used to a significant extent to form resist images. Perhaps more printed circuits have been made using screening than any other method. Thick, inexpensive, chemically resistant images can be formed with resonable accuracy at an exceptionally high production rate with screening. Once a screen is prepared, many boards can be screened in rapid succession, particularly with automatic equipment. Images are formed on tensioned screens with gelatin or polyvinyl alcohol (PVA) rendered photosensitive by dichromate or other means. Silk, stainless steel, nylon, and polyester screens are used. Several polymer-based screenable inks are available, typical of which are polyvinyls and polyacrylics.

The resist image is formed when a supply of ink is moved across the screen surface by the squeegee and forced through the screen opening onto the board. The squeegee depresses the screen against the board at a point of contact. As it passes a given point, the screen lifts and leaves ink forced through the openings. The squeegee blade forms a seal with the gelatin and forces the ink ahead of it, wiping clean the smooth surface. Screened resist images can be used with most processes for either pattern plating or etching and are applicable to most single- and double-sided board processes.

Dichromated resists have also been used in the manufacture of circuit boards. Naturally occurring colloid resists, such as gelatin and albumen, are generally unsatisfactory as they lack adequate chemical resistance, and they are unable to withstand a high enough postbake temperature to obtain it. The resists age poorly after sensitization and require unusual techniques; the coatings are affected by humidity. Further, many colloids are nutrients and are subject to molding or other forms of biologic degradation; so they must be used shortly after preparation for satisfactory results. Woods (68) indicates that some success in etching circuit boards can be obtained with bichromate gum arabic with concentrated ferric chloride (45° Baumé). The solubility of the exposed gum image is greatly reduced by the high chloride-salt concentration. However, the etch rate of these solutions is much lower. Fish-glue resists are widely used as an etch resist for copper engraving and halftone printing plates. After the image is formed, it is subjected to a high-temperature bake in the range of 350 to 550°F for a short time, during which it undergoes a "vitrifying" reaction, which renders the coating resistant to both acids and organic solvents. Any range of etchant concentration could be used, and the metal plates would be unaffected by the high-temperature postbake. Unfortunately, most circuit-board laminates would be damaged by such a bake. Dichromated shellac was perhaps the most successful of the naturally occurring materials for electronic applications. Extremely accurate images could be produced that were resistant to most acid solutions with a short postbake of less than 200°F. Because of its comparatively low processing temperature, shellac is sometimes referred to as "cold top," whereas the fish-glue resists requiring the high-temperature bake were quite naturally termed "hot-top" resists. Cold top is used today as an electroforming resist in the manufacture of precision storage-tube mesh (72). Line widths of 0.3 mil in 1,000-mesh screens are produced in large numbers daily. The resist is still quite sensitive to humidity changes and requires considerable technique to account for natural inconsistencies.

The second phase of the dichromate era in the evolution of photoresist technology was the introduction of synthetic photosensitive polymers. Polyvinyl alcohol (73–75), polyvinyl pyrrolidone (76), and polyvinyl butyral (77) were capable of sensitization with dichromates. These solutions had some of the expected advantages over their natural counterparts with regard to consistency and shelf life. Like other photoresists, they were initially used in the printing industry for etching and on lithographic plates. Polyvinyl alcohol, the best known and most widely used of the materials for circuit-board fabrication, was actually classified as a hot-top resist though the postbake temperature was considerably lower than that required for fish glue. Hardened PVA coatings could endure etching with a bake at about 450°F, which was within the allowable limits for some copper-clad laminates, with proper racking to reduce damage by warp and twist. The use of dichromate-sensitized PVA photoresist was mentioned by Eisler (8) in 1959 and as late as 1968 by Woods (68). In the 1960s, however, the primary use made of dichromate-sensitized PVA (or gelatin) has been in the preparation of screen stencil patterns for applying nonphotosensitive resists on circuit boards.

It was apparent that the processing problems and other shortcomings associated with the dichromated resists and screened patterns were inadequate for the total needs of the printed-circuit industry that was evolving in the late 1950s and early 1960s. Multilayer circuit boards were rapidly gaining popularity and were placing increasing demands on resist processes, particularly in the areas of accurate pattern repeatability and layer-to-layer pattern registration. KPR began gaining popularity in the late 1950s to fill the needs of the circuit-board industry (78,79), and its introduction marked the beginning of the modern era in the evolution of photoresists. KPR has nearly indefinite shelf life, and the coated parts can be stored for long periods of time before exposure and developing because they are not subject to the "dark reaction," which plagued the early resists (Chap. 2). Normal variations of temperature and humidity have much less effect on the processing parameters of KPR than the dichromated resists. The resist image is comparatively tough, has good chemical resistance, and can function either as a plating or etching resist. It was found that photoresist images were quite accurate (80), more so than screened images (Fig. 1-8), and many problems related to registration and alignment were eliminated. The data used to prepare Fig. 1-8 were gathered in 1961 by the author; subsequently, refinement in screening techniques has improved the accuracy, but screening techniques still do not approach the photoresist process. Similar results have been reported elsewhere (81).

Among the more significant developments to the industry was the introduction of the positive photoresists by the Shipley Company in the early 1960s. The resists, AZ-15 and AZ-17, do not have the overall chemical resistance of KPR but are extremely accurate, convenient to process (aqueous developer and no dyeing step), and readily removed. In 1962, Dynachem Corporation began what was to become the only complete line

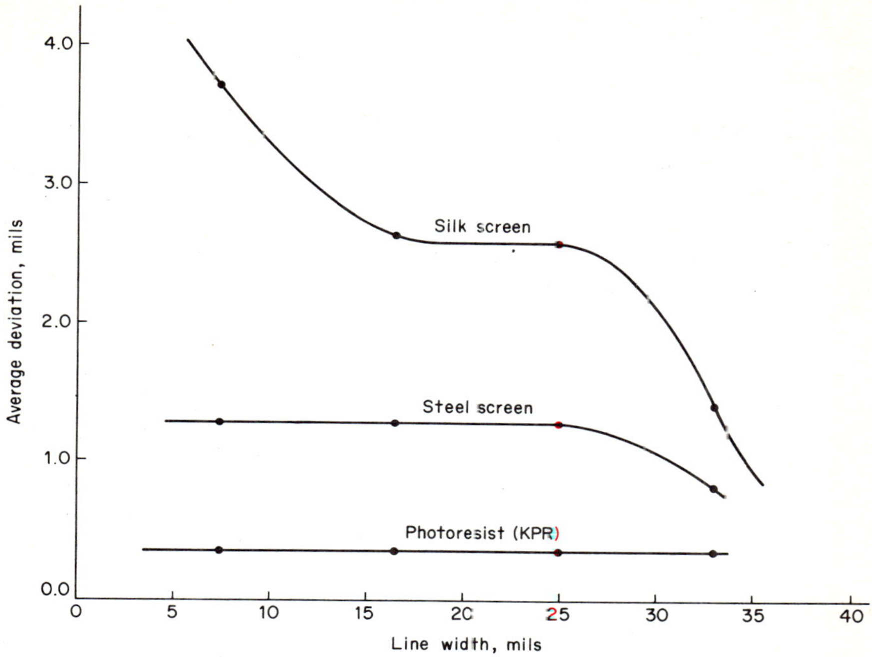

Fig. 1-8 Line-width deviations as a function of line widths. (*Autonetics Division of Rockwell International.*)

of photoresists offered by any company. P. A. Hunt Company introduced their line of resists in 1967, as did GAF (Table 1-1). Kodak, Shipley, and Dynachem continued to add to their growing list of photoresists.

Finally, the most recent breakthrough in photoresist technology was the introduction by E. I. du Pont de Nemours & Company in 1968 of the solid-film resist. The Du Pont Riston series of solid-film photoresists are sandwich structures in which the outer surface layer is a thin transparent Mylar film protecting the inner photosensitive coating layer until after exposure. This prevents dust and other contaminant particles from affecting the integrity of the film prior to and during exposure and eliminates oxygen effects. The overall result is a reduction or elimination of pinholes and other imperfections. Another advantage of solid-film resists is that they lend themselves better to automation because they are applied by a lamination process. The Meaker Company and the W. H. Brady Company invaded the expanding market near the end of the decade, the former with a series of wet-film resists and the latter with the Phodar series of dry films. Both have been discontinued, however.

MICROELECTRONICS

Besides their indispensable use in the manufacture of circuit boards, special-grade liquid photoresists were essential to the early development of transistors and other semiconductor devices. In the early work (1958) by the Diamond Ordnance Fuze Laboratories (82), photoresists were used to define the pnp or npn junctions. Today the entire semiconductor industry is dependent on the use of photoresists for the manufacture of their devices and circuits. Numerous photoetching steps are employed sequentially in order to form the various active and passive components of the circuit. Details of these processes will be described in a later section of this book.

SUMMARY

Generally, the evolution of photoresists can be divided into two eras—the dichromate era and the modern era. There have been many important discoveries, a few of which have been mentioned (Fig. 1-9).

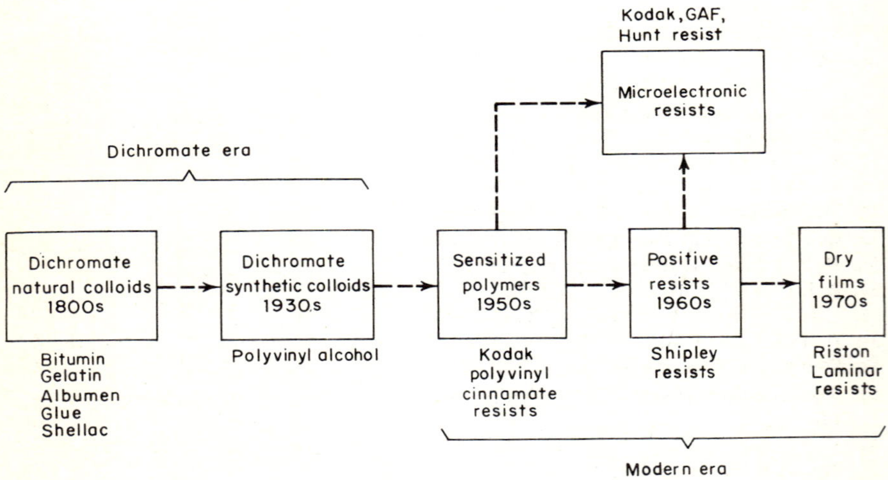

Fig. 1-9 Evolution of photoresists.

REFERENCES

1. G. W. Jorgensen and M. H. Bruno, "The Sensitivity of Bichromated Coatings," Lithographic Tech. Foundation, New York, 1954, *Bull. 218*, p. 1.
2. R. S. Cox and H. L. Howard, *Process Eng. Mon.*, vol. 66, pp. 10–13, 1959.
3. J. Kosar, *Light Sensitive Systems*, John Wiley & Sons, Inc., New York, 1965, p. 194.
4. P. Glafkides, *Photographic Chemistry*, Fountain Press, London, 1960, vol. 2, p. 676.
5. A. Murry, U.S. Patent 1,965,710 (1934).

6. C. C. Unruh and C. F. H. Allen, U.S. Patent 2,716,102 (1955).
7. L. M. Minsk, U.S. Patent 2,725,372 (1955).
8. P. Eisler, *The Technology of Printed Circuits*, Heywood and Company, Ltd., London, 1959, p. 103.
9. R. S. Cox and H. L. Howard, *Process Eng. Mon.*, vol. 66, pp. 141–142, 1959.
10. T. D. Schlabach and D. K. Rider, *Printed and Integrated Circuitry*, McGraw-Hill Book Company, New York, 1963, pp. 85–86.
11. C. Ducas, U.S. Patent 1,563,731 (1925).
12. P. Eisler, *op. cit.*, pp. 3–6.
13. P. Eisler, U.S. Patent 2,441,960 (1948). Also see U.S. Patent 2,587,568 (1952).
14. P. Eisler, U.S. Patent 2,607,825 (1952).
15. P. Eisler, U.S. Patent 2,748,071 (1956).
16. P. Eisler, U.S. Patent 2,747,977 (1956).
17. J. T. Beck, U.S. Patent 2,823,286 (1958).
18. J. B. Little and B. E. Phelps, U.S. Patent 2,902,629 (1959).
19. O. I. Steigerwalt and H. I. Ashry, U.S. Patent 2,912,745 (1959).
20. H. I. Ashry, O. I. Steigerwalt, and J. D. Heibel, U.S. Patent 2,912,746 (1959).
21. H. I. Ashry, O. I. Steigerwalt, and J. D. Heibel, U.S. Patent 2,912,747 (1959).
22. W. D. Frazier, U.S. Patent 2,915,678 (1959).
23. R. B. Gray, U.S. Patent 2,912,748 (1959).
24. N. N. Berger, U.S. Patent 2,889,393 (1959).
25. H. Shapiro, *Plating*, June, 1957, pp. 607–611.
26. D. K. Cardy, U.S. Patent 2,864,156 (1958).
27. R. C. Sanders, U.S. Patent 2,876,392 (1959).
28. M. Martin and C. S. Bojanowski, U.S. Patent 2,861,911 (1958).
29. S. K. Tally and V. F. Dahlgren, U.S. Patent 2,876,393 (1959).
30. W. G. Lee, U.S. Patent 2,940,018 (1960).
31. J. F. Dytrt, *Ind. Eng. Chem.*, March, 1959, pp. 286–287.
32. S. Wein, *Metal Finishing*, August, 1948, p. 58.
33. H. Narcus, U.S. Patent 2,421,079 (1947).
34. H. Narcus, U.S. Patent 2,454,610 (1948).
35. Q. Marino, U.S. Patent 1,452,281 (1923).
36. B. F. Walker, U.S. Patent 2,303,871 (1942).
37. J. J. Gitto, U.S. Patent 2,897,409 (1959).
38. J. H. Hauser, U.S. Patent 2,972,391 (1959).
39. P. Talmey, U.S. Patent 2,848,359 (1958).
40. W. G. Lee, U.S. Patent 2,940,018 (1960).
41. *Technical Data Sheet PC-101*, MacDermid Incorporated, Waterbury, Conn., Apr. 13, 1967, p. 2.
42. E. H. Babcock, R. K. Stephens, and R. C. Grimsinger, *43rd Annual Tech. Proc.*, Amercian Electroplaters' Society, 1956, pp. 267–271.
43. W. J. Hodges, "Eyelet Failure in Etched Wiring," Hughes Aircraft Company, Communication Systems Laboratories, May 5, 1957.
44. B. C. Alzua, Jr., *Missiles and Rockets*, vol. 5, pp. 37, 38, July 27, 1959.
45. R. B. Atkinson, U.S. Patent 3,119,709 (1964).
46. M. C. Agens, U.S. Patent 2,938,805 (1960).
47. R. H. Thielemann and D. L. Benedict, "Research on Printed Circuitry," *Stanford Research Institute Project S-1997, Tech. Rep. 1*, December, 1956.
48. E. B. Saubestre, *46th Annual Tech. Proc.*, American Electroplaters' Society, 1959, pp. 264–276.
49. E. B. Saubestre, *Sylvania Technol.*, vol. 12, no. 1, pp. 6–11, 1959.
50. R. M. Lukes, U.S. Patent 2,996,408 (1961).

51. F. W. Schneble, U.S. Patent 3,033,703 (1962).
52. "How to Bond to Copper while Metallizing Plastics," Shipley Company, *Data Sheet DS–12A*, January, 1960.
53. J. F. Jones, D. F. Bowman, and C. M. Dowler, "Methods and Techniques for the Production of Printed Circuit Boards," *Autonectics Tech. Rep. EM–1363*, Aug. 29, 1957.
54. C. R. Shipley, Jr., U.S. Patent 3,011,920 (1961).
55. R. Chan, U.S. Patent 2,990,310 (1961).
56. R. O. Stoehr, U.S. Patent 3,028,573 (1962).
57. "Military Specification for Printed Wiring Boards," *MIL-P-55110A*.
58. "Printed Wiring for Electronic Equipment," *MIL-STD-275A*.
59. "Acceptability of Printed Circuits," Institute of Printed Circuits, October, 1964.
60. "Plated Through Hole Multilayer Circuit Boards," *MIL-P-55640*. Also see *MIL-STD-1495*.
61. D. E. Yost, *Symposium on Printed Circuits*, University of Pennsylvania, Jan. 20, 1955, pp. 53–56.
62. S. W. Chaikin, J. Janney, F. M. Church, and C. W. McClelland, *Ind. Eng. Chem.*, vol. 51, no. 3, pp. 299–304, March, 1959.
63. E. F. Harmon, "Fabrication of Multilayer Boards at Autonetics for Minuteman II Program," *NAR-Autonetics Publ. X5-562/33*, March, 1965.
64. R. A. Nagy, U.S. Patent 3,357,099 (1967).
65. T. D. Nierenberg, U. S. Patent 2,602,731 (1952).
66. *Shipley Process Instruction P1340*, Shipley Company, Newton, Mass.
67. *Du Pont Riston Tech. Manual*, plating and etching section, p. 44.
68. B. Woods, *Trans. Inst. Metal Finish.*, *Printed Circuit Suppl.*, 1968, pp. 131–136.
69. Author's personal experiences.
70. F. M. Gillett and A. A. Katz, U.S. Patent 2,965,952 (1960).
71. "Research on New Approaches to Printed Circuitry," technical staff, Haloid Company, *ASTI Doc. AD-103234*, June, 1956.
72. Visit with Ray Grossman of Buckbee-Mears, St. Paul, Minn., July, 1969.
73. C. A. Brown, U.S. Patent 2,742,358 (1956).
74. C. A. Brown, U.S. Patent 2,830,899 (1958).
75. A. F. Greiner, U.S. Patent, 2,174,629 (1939).
76. F. F. Frost, U.S. Patent 2,624,673 (1953).
77. T. T. Boersma, U.S. Patent 2,819,164 (1958).
78. M. P. Forte and H. Shapiro, "Technical Note RADC-TN-56-198," Rome Air Development Center, USAF, *ASTI Doc. AD-103234*, June, 1956.
79. W. P. van Deusen, *Plating*, February, 1958, pp. 151–160.
80. J. R. Sayers and J. Smit, *Plating*, July, 1961, pp. 789–793.
81. J. W. Dini and A. D. Andrade, *Metal Finishing*, February, 1966, pp. 46–51.
82. J. R. Nall and J. W. Lathrop, "The Use of Photolithographic Techniques in Transistor Fabrication," *Diamond Ordnance Fuze Laboratories Rep. TR-608*, *ASTI Doc. AD-159233*, June 1, 1958.

CHAPTER TWO

Types of Photoresists

For many of us, the dark brown bottles of photoresist we have bought and used for many years were and perhaps still are pretty much a mystery. We were familiar with their properties through experience, but information as to why something happened was scarce, to say the least. Upon reflection, it is not strange that formulators were reluctant to release such information because of the many potential competitors becoming interested in these products. Photoresist formulations were and are highly proprietary, and the field is quite competitive. Whenever an innovation was marketed, competition quickly developed.

Formulators were faced with a very basic problem—they could not afford to market a resist to fit everyone's needs. They had to formulate for the bulk of the market, which was beginning to diverge with the serious growth of the microelectronics industry. As a result, specific needs, particularly with regard to purity and repeatability, were filled by the users themselves. Gradually, information regarding the basic composition of photoresists became available in literature, some of which has been compiled herein for the edification of the reader. In addition to the presentation of general composition information about the positive and negative photoresists, it is a specific aim to relate the behavior of the resists which we have all most likely observed to the basic composition. It is further hoped that the information in this chapter will be useful in the development of processing innovations to meet unusual needs and that it may aid in the selection of photoresists. It should be pointed out that an understanding of this chapter is not a necessary prerequisite to use the other portions of this book. However, every effort has been made to present a complex subject in a simplified form in this chapter.

It may be observed that photoresists contain at least two components—(1) a film-forming resin dissolved in (2) a solvent system. Further observations will confirm that the dried film undergoes changes when exposed to light radiation of a certain wavelength; this exposure imparts greater or less chemical resistance to the exposed areas in comparison with unexposed areas. If the resist is negative-acting, the exposed areas are hardened, whereas positive resists are solubilized by exposure. It is apparent, then, that the changes in the exposed areas are effected by some chemical reaction due to the absorbed radiation. Historically, it is possible that the resin itself is intrinsically photosensitive (e.g., Syrian asphalt), but a much more likely conclusion would be that a substance or substances which are capable of absorbing light energy and bringing about an insolubilizing reaction (e.g., ammonium dichromate) have been added to the formulation. These substances are generally referred to as "sensitizers" or "photoinitiators" and are invariably present in modern-day photoresist formulations. This brings the total constituents to three. A fourth group, which has simply been categorized as "additives," is also present. In this group are substances which enhance or improve the performance of the resist. Examples include antioxidants, stabilizers, and related materials. In the remaining portions of this chapter, these four categories will be discussed for each type of resist.

NEGATIVE RESISTS

Negative-acting resists of the modern era (since 1950) characteristically have high chemical resistance and good image reproduction qualities and are of low cost. They have been and are widely used in the manufacture of circuit boards and microelectronic devices for these reasons. Because of their high chemical resistance, the negative-acting resists are generally more difficult to remove than other resists although there are satisfactory commercial strippers. Negative-resist constituents are considered individually in the following discussion.

Resins

The resin system used in negative photoresists does not have to be photosensitive, but it does have to be capable of insolubilization through a reaction with the sensitizer molecule, which has been activated by absorbed light energy. Some other basic properties of negative-resist resins are listed in Table 2-1.

Of obvious interest is the method by which the resist is rendered insoluble to the developing solution upon exposure. With the exception of the dichromated resists, one characteristic that the modern-day photoresist resins possess is "chemical unsaturation." This term in organic chemistry refers to a chemical bond between two carbon atoms through two shared electron pairs in the same molecule. Upon activation, the one electron

TABLE 2-1 Basic Properties of Negative-acting Resist Resins

1. The resin must be capable of being insolubilized upon exposure to actinic light.
2. Prior to exposure, the resin should be soluble in a solvent system.
3. The resin should be capable of wetting and adhering to a variety of surfaces.
4. Upon air evaporation of the solvent system, the resin must form a uniform coherent film without losing its solubility in the solvent.
5. The processed resin image must be resistant to aqueous solutions.

pair can be shared by carbon atoms of adjacent polymer molecules which are similarly unsaturated. The sharing of an electron pair by carbon atoms in adjacent polymer molecules is referred to as "cross linking" and has a profound effect on the physical properties of coatings in which it occurs. Cross linking increases the hardness, chemical resistance, and the adhesion to the substrate.

"Polymers" are groups of molecules that contain repetitive sequences of atoms; they are noted for their high molecular weight and large size. Most polymers that make up photoresist resins are very large chainlike structures that contain unsaturated linkages; they are classified as linear polymers. The basic shapes of polymer molecules are shown in Fig. 2-1. Linear polymers contain a basic backbone structure, which may have various functional groups connected to them. Some linear polymers that are of concern are from a group called "vinyl" or "polyvinyl" compounds. The former term is applied to the monomer or building blocks from which the larger polyvinyl polymer is formed (Fig. 2-2). One of the most widely used plastics for tank liners and chemical etcher fabrication is polyvinyl chloride, PVC. Other notable examples are polyvinyl benzene (styrene) and polyvinyl alcohol.

Linear polymers in a film are attached only loosely to each other by secondary forces which are weaker than electron bonds. Upon sufficient heating, thermal energy is absorbed, the secondary forces are broken, and the molecules move easily over and around each other. The resulting physical change is eventually a return to the liquid state, or, simply stated, melting occurs. Polymers which melt in this manner are referred to as

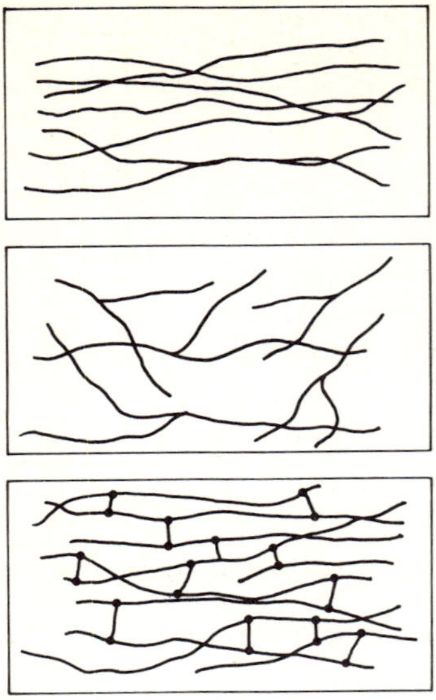

Fig. 2-1 Forms of polymers. (*a*) Linear polymers—polyvinyl cinnamate; (*b*) branched polymers (not interconnected); (*c*) cross-linked polymers (interconnected, sometimes called "network" polymers; virtually one great molecule).

"thermoplastics." Branched polymers behave generally as the linear polymers. If the functional groups on the vinyl backbone contain unsaturated groups, heat or thermal polymerization may take place before melting occurs. The result is a polymer that is a highly cross-linked network similar to Fig. 2-1*c*. This cross-linked, or network, polymer becomes quite different from the linear polymers from which it was formed. As indicated previously, it is generally stronger and more brittle, it possesses greater thermal and chemical resistance, and it does not heat-soften. The interconnected structure of the cross-linked, or network, polymer prevents the molecules from moving with the freedom of the linear polymer when thermal energy is absorbed. Instead of melting, the cross-linked, or network, polymer remains a solid until enough energy is absorbed to cause the chemical bonds to rupture and form different lower-molecular-weight compounds. Eventually, charring and total decomposition will occur with prolonged heating at high temperature. Polymers of this type are said to be "thermosetting." Photoresist coatings are similar to a thermoset resin after exposure to actinic radiation (light). In lieu of thermal energy, photochemical energy is responsible for the cross-linking reactions. Further polymerization can be brought about by postexposure baking of the image.

When the exposed part is immersed in the developing solution, areas that have not been exposed retain their original linear (or branched)

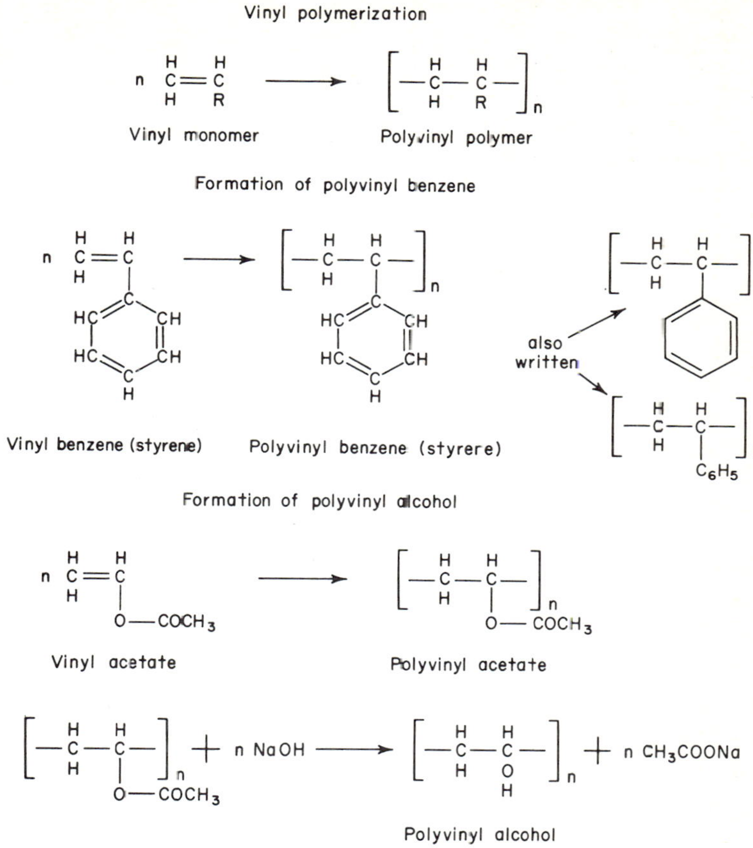

Fig. 2-2 Some vinyl polymers.

solvency and are removed by solvent action. Properly exposed areas have been polymerized to a point where they have lost their solvency and are able to resist the developer action. Thus, a negative image is formed by the selective solvency of the exposed and unexposed areas in the developer.

In addition to the unsaturation and solvent solubility, the resin must be adherent to metal surfaces and capable of wetting them thoroughly. Wetting occurs when the polymer molecules have a strong attraction for the metal surface. Poor wetting occurs when the polymer molecules have strong attractive forces between themselves which reduce the attraction for the surface. Adhesion is the result of an attraction between sites on the metal surface and the polymer molecule (although there are several theories on the subject).

While the resin must be capable of photopolymerization and preferably some thermal polymerization, it should not be polymerized significantly by air, as paints and varnishes when the solvent system evaporates. It should, in fact, be resistant to polymerization of this type to such an extent that moderate heating in air can be tolerated (preexposure baking). From practical experience with prebaking, some resins (or sensitizers) are more resistant than others to thermal polymerization.

Some information is available regarding the resin systems used in the more widely known photoresists. Table 2-2 lists those resins in this category.

TABLE 2-2 Some Basic Negative-acting Photoresist Resins (Liquid Systems)

Resin	Probable resists
Polyvinyl cinnamate derivatives	KPR, KPR-2, KPR-3, and KPR-4; Waycoat RC and 20
Vinyl ester–containing cinnamylidene	KOR
Allyl ester prepolymer	DCR wet-film resists
Cyclized-rubber derivatives	KMER, KTFR; Waycoat IC, SC, and 450
Hydrocarbon copolymer	Dynachem CMR series

Polyvinyl Cinnamate. Starting in 1951, L. M. Minsk and his coworkers patented (1–5) processes for preparing light-sensitive compositions containing unsaturated esters of polyvinyl alcohol, among which was polyvinyl cinnamate. The resin image, after exposure and developing, functions as the ink-receptive surface on lithographic plates (see Fig. 1-7). Several preparation methods were tried that proved unsuccessful, including polymerization from the vinyl cinnamate monomer with benzoyl peroxide (Fig. 2-3). Apparently, some cross linking occurred between the unsaturated vinyl and cinnamoyl groups, which yielded a thermoplastic that was neither solvent-soluble nor photosensitive. The successful method utilized a derivative of cinnamic acid, cinnamoyl chloride, in which the —OH portion of the carboxyl group (acid) is replaced by a chloride. When reacted with polyvinyl alcohol, the chloride combined with the hydrogen on the alcohol to form hydrogen chloride. The oxygen atom on the alcohol and the carbon atom containing the carbonyl group also combined as shown in Fig. 2-3. The reaction took place in a solvent which had an affinity for the hydrogen chloride, which prevented it from recombining with the reactants (pyridine). When the reactants combined fully after heating several hours, the solution was diluted with acetone to dissolve the polyvinyl cinnamate which had precipitated. Most of the

$$(1)\quad R{-}COOH + R'{-}OH \longrightarrow R{-}COOR' + H_2O$$

Acid Alcohol Ester

$$(2)\quad R{-}COOH + HCl \longrightarrow R{-}COCl + H_2O$$

$$R{-}COCl + R'{-}OH \longrightarrow R{-}COOR' + HCl$$

Ester

Vinyl cinnamate

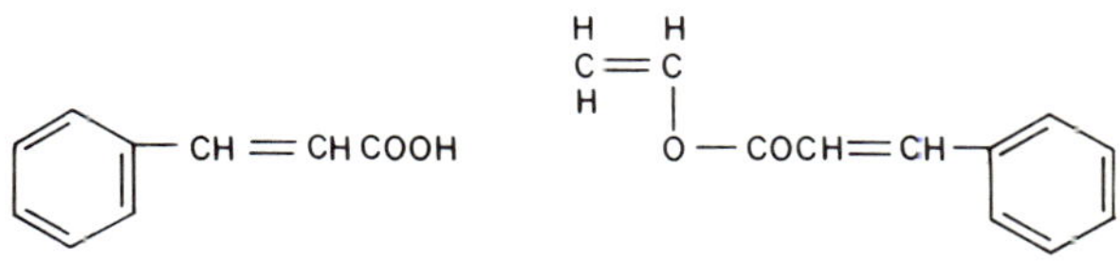

Cinnamic acid Vinyl cinnamate ester

Vinyl cinnamate —(Polymerization / Benzoyl peroxide)→ Solid, insoluble, highly cross-linked mass. Not photosensitive

Polyvinyl cinnamate

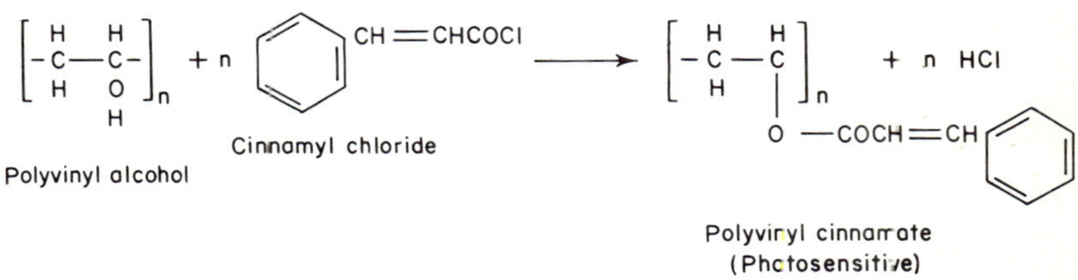

Fig. 2-3 Polyvinyl cinnamate preparation.

reaction products could be removed by washing with water, leaving the precipitated resin. After it was washed thoroughly, the resin was dried and dissolved in ethylene glycol monoethyl ether acetate (Cellosolve acetate). To ensure full esterification of the alcohol groups, an excess of cinnamoyl chloride was added. For each pound of polyvinyl alcohol, about 4.5 lb of cinnamoyl chloride was used to ensure chemically that an excess was present. The yield was about 2.8 lb of high-purity polyvinyl cinnamate that was virtually free of unreacted alcohol groups.

The polymer produced by this method was intrinsically photosensitive and solvent-soluble. When dissolved in Cellosolve acetate to form a dilute (2.5 to 3.5 wt percent) solution and a sensitizer added, a suitable coating solution for lithographic plates could be obtained. The solution was flowed onto zinc lithographic plates and whirled dry. Following exposure, the image was developed by swabbing with Cellosolve acetate or by using trichlorethylene vapor. Polymer images thus formed were reportedly ink-receptive, accurate, tough, and adherent.

The physical properties of polyvinyl cinnamate polymers can reportedly be varied substantially by one of two methods. The degree of esterification can be varied over a range of 60 to 100 percent (Fig. 2-4), and the length of the vinyl chain can also be controlled. Partial esterification simply means that not all the available alcohol groups are reacted with cinnamoyl groups, as shown in Fig. 2-4. This may be accomplished by one of two

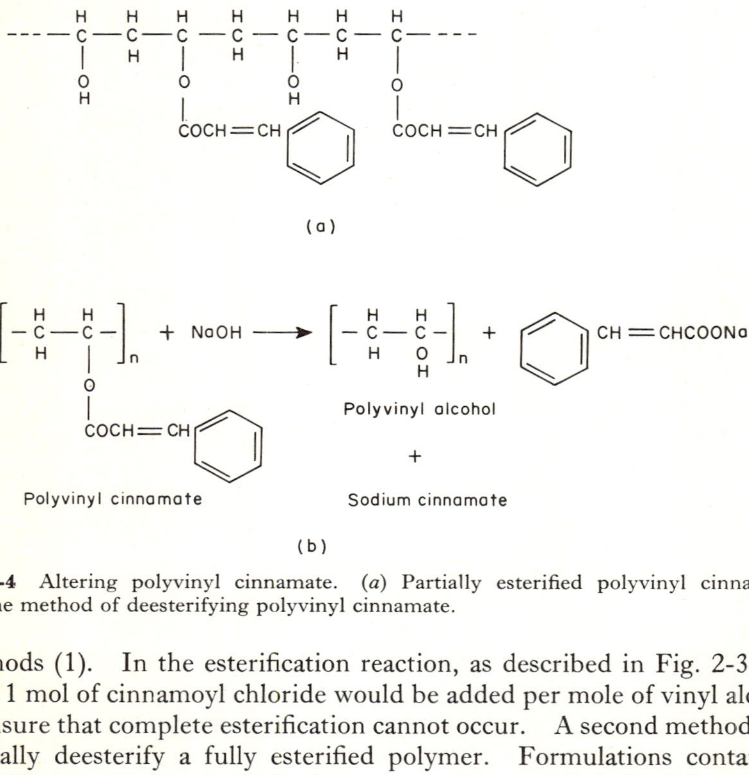

Fig. 2-4 Altering polyvinyl cinnamate. (*a*) Partially esterified polyvinyl cinnamate; (*b*) one method of deesterifying polyvinyl cinnamate.

methods (1). In the esterification reaction, as described in Fig. 2-3, less than 1 mol of cinnamoyl chloride would be added per mole of vinyl alcohol to ensure that complete esterification cannot occur. A second method is to partially deesterify a fully esterified polymer. Formulations containing partially esterified polyvinyl cinnamate were patented by Murry and Leubner (6). Fully esterified polyvinyl cinnamate is mixed with the partially esterified resin. If the degree of esterification falls below about 55 mol percent (approximately 55 percent of the available hydroxyl groups are esterified), the resulting resin is difficult to dissolve in a common solvent; this in turn causes poor developing and scumming.

KPR, KPR-2, and Waycoat RC appear to contain fully esterified polyvinyl cinnamate. In an analysis of KPR-2 that was published by

Levine, Lesoine, and Offenbach (7), it was shown that the resin was fully esterified. Infrared spectra will show little hydroxyl bond absorption on any of these resists.

The photosensitivity of polyvinyl cinnamate is due to the reactivity of the cinnamoyl group. Cinnamic acid is a naturally occurring compound (found in oil of cinnamon, from which it derives its name) and is readily synthesized. Its photosensitivity has long been known, though little use was made of this property until fairly recently. In its solid state, cinnamic acid undergoes a reaction known as "dimerization," in which two molecules combine (Fig. 2-5*a*). Other derivatives, including esters, undergo this reaction (8), including polyvinyl cinnamate (9,10). Figure 2-5*b* shows a probable method by which cross linking occurs upon exposure (9).

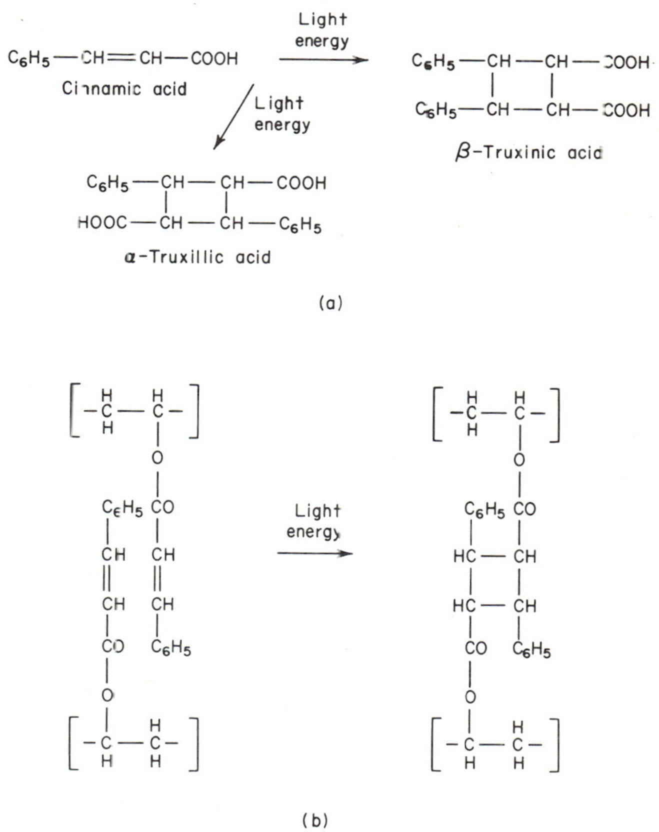

Fig. 2-5 Photochemical reactions of the cinnamate. (*a*) Some photochemical reactions of cinnamic acid (dimerization); (*b*) cross linking of polyvinyl cinnamate by α-truxillic dimerization.

Murry (11) attempted to utilize the photosensitive properties of cinnamic acid derivatives (cinnamal ketone) in the early 1930s to prepare an ink-receptive imaging solution for lithographic plates. He observed that in the crystalline state cinnamic acid undergoes the reactions shown in Fig. 2-5*a* rather slowly and that they are of little value in forming films. To prevent crystallization, he mixed the acid in a common solvent with a nonreactive film-forming resin. Apparently, some polymerization occurred under these conditions upon exposure to ultraviolet light, and the exposed areas were rendered insoluble and could resist developing. In the late 1940s, Allen and Van Allen (12) succeeded in preparing a photosensitive vinyl cinnamoyl polymer from styrene and cinnamoyl chloride. This polymer and some others (12–16) (Fig. 2-6) were prepared in the early and mid-1950s.

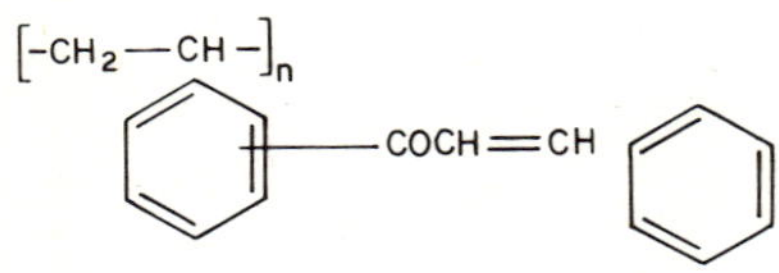

Polystyrene - cinnamyl ketone

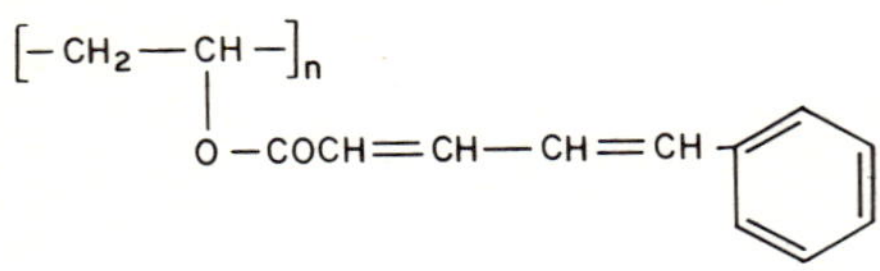

Polyvinyl cinnamylidene ester

Fig. 2-6 Photosensitive resins related to cinnamic acid. KOR is reported to contain a resin similar to the ester.

It is interesting to note the spreading use of photosensitive compositions containing polyvinyl cinnamate. From Japanese studies, it is probable that the resin is used in Japan (10,17). Polyvinyl cinnamate has also been used in microelectronics (18).

Some attempts were made (2–5) to utilize cellulose as the backbone chain in lieu of polyvinyl alcohol. Cellulose has a repeating unit that contains three hydroxyl groups which can be esterified. Early use was made of this property by esterification to form cellulose acetate and nitrate esters. A preference for polyvinyl cinnamate in lieu of cellulose cinnamate was indicated in the references, however.

Kodak ortho resist (KOR), the most photosensitive of Kodak resists, has not been widely publicized, and information regarding the resin system is quite limited. Collins and Deverse (19) indicated that the resin was a cinnamylidene acrylic acid ester. Unusually high photoreactivity is afforded by the conjugate double bonds (—CH=CH—CH=CH—) (Fig. 2-6), particularly with a carbonyl (—C=O) unit adjacent to the group.

Allyl Ester Resins. The first and apparently only liquid photoresist preparation of commercial significance utilizing allyl esters (Fig. 2-7) is Dynachem photoresists (DCR). At a technical society meeting (20), Martinson, one of the inventors of DCR, and later Herzog, indicated that the resin system was a diallyl phthalate derivative. From the examples in the patent (21), diallyl isophthalate may be preferable. Work on DCR apparently started before 1963, when the patent was filed. Other allyl

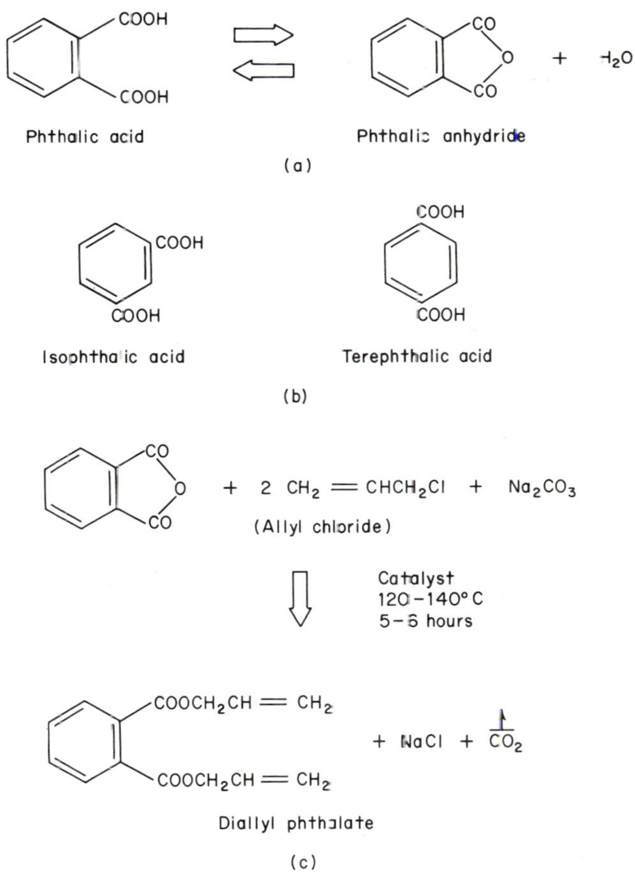

Fig. 2-7 Diallyl phthalate. (*a*) Phthalic acid and anhydride; (*b*) isomers of phthalic acid; (*c*) preparation of diallyl phthalate.

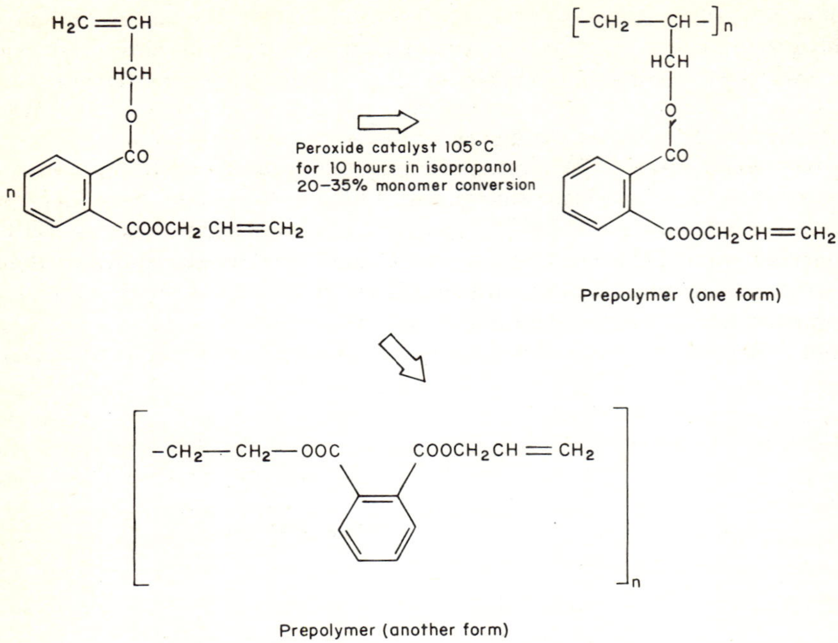

Fig. 2-8 Diallyl phthalate prepolymer resin.

esters (22) were also patented by the inventors of DCR (Fig. 2-8). Unlike polyvinyl cinnamate, this resist was not originally formulated for lithographic purposes but rather for printed circuit boards and related electronic applications. A lithographic formulation was subsequently prepared (23), however.

Diallyl phthalate and isophthalate are products of commerce (24,25). The resin is prepared from the monomer by partial polymerization (26,27) (Fig. 2-8). The monomer is prepared from allyl chloride and the acid anhydride (28). Polymerization is carefully controlled and stopped when about 20 to 35 percent of the monomer is polymerized. The partially polymerized thermoplastic resin thus formed is referred to as the "prepolymer." If the polymerization is not controlled and allowed to continue past the prepolymer or beta (β) stage to the gamma (γ) stage, a highly polymerized (cross-linked) thermosetting plastic resistant to acids, alkalies, and solvents is formed. The prepolymer contains a residual unsaturation and is frequently employed for precision molding (with a catalyst) because the shrinkage between the β and γ stage is only about 1 percent.

The prepolymer is a white powder that is soluble in a variety of solvents, such as acetone and Cellosolve acetate, and possesses good dimensional stability, physical strength, and high-temperature resistance. Kodak also patented (29) a photosensitive composition utilizing a diallyl phthalate (or isophthalate) prepolymer resin system. The composition was reportedly useful as both an etching resist and for lithographic purposes.

Other photosensitive resins utilize phthalic acid esters (30–33); however, systems utilizing allyl unsaturation are strangely missing.

Isoprenoid Resins. The metal-etch-type resists such as Kodak metal etch resist (KMER) (34), Kodak thin film resists (KTFR) (7), Waycoat IC and SC (35), and Dynachem CMR and Micro-G (35) series reportedly utilize resin systems that contain derivatives of polyisoprene rubber. However, Dynachem indicated that CMR was a hydrocarbon copolymer. Polyisoprene is the basic polymeric substance in natural rubber, where its molecular weight can reach a million or more. Synthetic polyisoprene can readily be prepared and is a commercial product. Both natural rubber and polyisoprene have a common structure and behave similarly chemically. Figure 2-9 shows the acknowledged chemical structure of the rubber. As may be noted, it contains one double bond for each five carbon atoms and contains only carbon and hydrogen atoms in its molecules. However, neither the natural nor the synthetic rubber is particularly suitable for use as a photoresist resin until it undergoes a process termed "cyclization." In this process, the rubber is physically changed

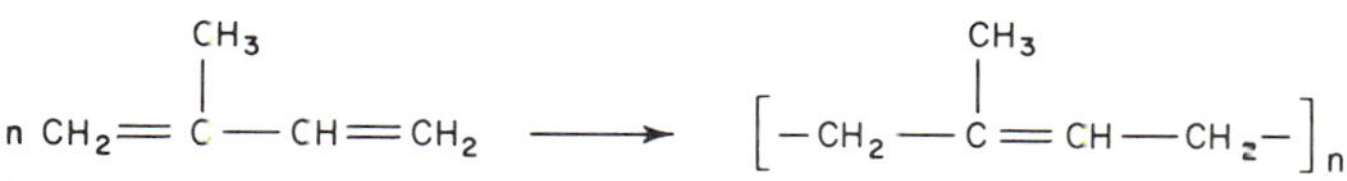

Preparation of synthetic cis-1,4-polyisoprene

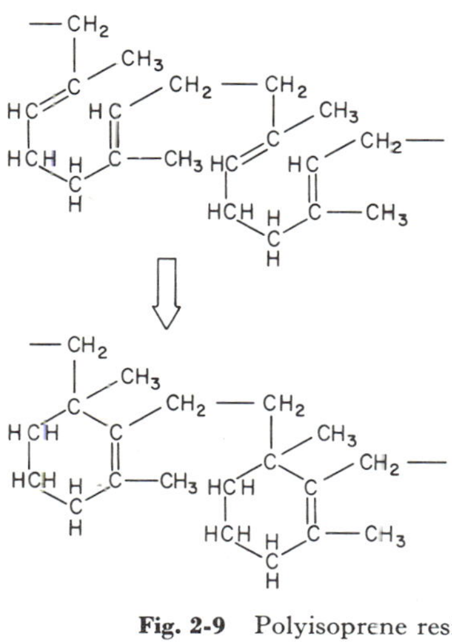

Rubber (natural or synthetic)

Repeating structure of the polyisoprene polymer. Note that there is one double bond for each five carbon atoms.

Resin

Isomeric after structure monacyclization. (D'Ianni, Reference 38) Note that there is only one double bond for each ten carbon atoms.

Fig. 2-9 Polyisoprene resin formation.

from its characteristic state to a tough, resinous material that is resistant to both acids and alkalies and adheres well to metal surfaces. As a process, cyclization has long been known, and cyclized rubber was available commercially before World War II (see Fig. 2-9). Uses included stiffeners for rubber to obtain better flow and molding characteristics; coatings for metals; as an electrical insulation; as an adhesive; and in paints, inks, and varnishes.

As with other photoresists, cyclized rubber was introduced initially by Kodak both in the United States (36) and Great Britain (37) as an ink-receptive photosensitive composition primarily for lithographic plates. Considerable effort was devoted to ascertaining the structure of the cyclized rubber, with that proposed by D'Ianni and his coworkers (38) in 1946 being generally accepted. More recently, Agnihotri and his coworkers

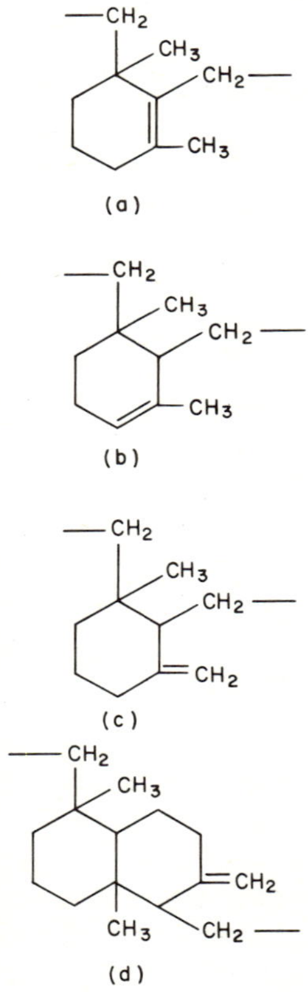

Fig. 2-10 Cyclization modes. Isomeric monocyclic structures of cyclized polyisoprene. The structure (*a*) proposed by D'Ianni and isomers (*b* and *c*) all contain one double bond for every ten carbon atoms. The bicyclic structure (*d*) contains one double bond for every fifteen carbon atoms. Other multicyclic structures are also possible.

(39), through the use of advanced analytic techniques including high-temperature nuclear-magnetic-resonance studies, indicated four cyclization structures can be obtained in the preparation of resins for photoresist. These structures are shown in Fig. 2-10. It was indicated that the degree of cyclization and the polymer size were factors that affected the solubility, image-formation ability, and resistance to etchants. Complete cyclization is difficult if not impossible to achieve. Some C_5H_8 groupings are always present in varying amounts (40).

From the structure of D'Ianni, it is expected that one unsaturated grouping occurs for every $C_{10}H_{16}$ combination. If it is further assumed that this structure repeats throughout the polymer, the level of unsaturation can be conveniently expressed as a function of the moles of unsaturation per the molecular weight of the repeating unit. For example, the unsaturation in the D'Ianni structure expressed in this manner would be 1 mol/$C_{10}H_{16}$, or 1 mol/136 g or 0.00735 mol/g.

TABLE 2-3 Polyisoprene Cyclization and Unsaturation

Polymer	Repeating unsaturate unit: Structure	Repeating unsaturate unit: Mol (g)	Level of unsaturation (mmol/g)
Natural rubber (uncyclized)	C_5H_8	68	14.7
Monocyclic structure (D'Ianni and isomers)	$C_{10}H_{16}$	136	7.4
Bicyclic structure	$C_{15}H_{25}$	205	4.9
Flory's structure (86.5% cyclization) (40)	Mixture of C_5H_8 and $C_{10}H_{16}$	. . .	8.4*
Optimum structure (about 90% bicyclic plus 10% C_5H_8 groups) (39)	. . . .	. . .	5.9

* Using 86.5 percent cyclization to bicyclic rubber, a level of unsaturation of 6.2 is obtained.

As a matter of convenience, "millimole" (1/1000 mole) is used in lieu of "mole"; this gives 7.35 mmol/g as the level of unsaturation. The level of unsaturation for other structures is shown in Table 2-3. From the data of Agnihotri, an unsaturation level of close to 6 mmol/g will produce a highly acceptable isoprenoid photoresist resin. This level was achieved by controlling the degree of cyclization to allow a combination of about 90 percent bicyclic structure and 10 percent uncyclized groups.

The molecular weight of the polymer is also a factor in obtaining the desired results from the resin. In natural rubber, molecular weights exceeding a million occur frequently. High-molecular-weight polymers in this range tend to produce poor images and because of their limited solubility form only dilute solutions (<10 percent). Cyclization tends to reduce the molecular weight of the rubber polymers by chain scission, apparently. KTFR resin has a reported (7,19) average molecular weight of approximately 60,000 while KMER is lower.

There are four basic methods used to cyclize rubber according to Nauton (41):

1. *Heat*—250 to 300°C in a solvent or inert gas
2. *Dehydrohalogenation of a rubber hydrochloride*—removal of chlorides from a rubber hydrochloride in an alkaline solution followed by reduction with zinc dust and acid
3. *The action of amphoteric metal halides*
4. *The action of acidic reagents*

Only the last two methods have achieved commercial significance and may have been used for the preparation of photoresist resins. In Method 3, amphoteric metal-halide salts, such as ferric chloride, boron trifluoride, aluminum chloride, and stannic chloride, particularly chlorostannic acid dihydrate ($H_2SnCl_4 \cdot 2H_2O$), act as catalysts for cyclization (42). The catalyst (8 to 10 wt percent of the rubber) is mixed with masticated (shredded) rubber dissolved in a suitable solvent, such as benzene, and reacted at about 75°C for 90 min. Upon completion of the reaction, water is added to destroy the catalyst by hydrolysis, and the resin is precipitated. After it is washed thoroughly, the resin is dried in a vacuum to yield an amorphous, lightly colored thermoplastic material. This method is applicable to both natural rubber and synthetic polyisoprene, with similar reaction conditions to obtain resins with similar softening points (65 to 90°C).

Resins prepared by this method usually contain an ash (about 1 percent) of stannic oxide (SnO_2) or metastannic acid (H_2SnO_3) formed by the hydrolysis of the catalyst; this is difficult to remove completely. A small amount of combined chlorine is also presented in the resin.

Acidic reagents (Method 4) have long been used to cyclize rubber. Examples (43) include sulfuric acid, *p*-toluene sulfonic acid, *p*-toluene sulfonyl chloride, and phenol sulfonic acid derivatives. Both natural and synthetic rubber are cyclized by these catalysts.

In the studies of unsaturation levels, Agnihotri (39) referenced the methods of Janssen (44) in the preparation of cyclized rubber from synthetic polyisoprene. The rubber is mixed with 5 to 15 pbw *p*-toluene sulfonic acid in a solvent such as toluene and refluxed. Cyclization is an exothermic reaction, so a temperature rise occurs (180°C). The reaction is stopped by neutralization of the acid, and the resin is precipitated, washed, and dried. An accelerator can be added to produce highly cyclized resins. Large amounts of the sulfonic acid catalyst and longer reaction times have a similar effect.

Cyclized natural rubber can be prepared directly from *Hevea* latex using the method developed by Gordon (45,46). *Hevea* latex is the milky liquid obtained by tapping the *Hevea brasiliensis* tree. It contains about 35 percent rubber dispersed in a neutral aqueous solution. The rubber particles dispersed in the liquid phase are negatively charged, and the addition of acids tends to neutralize the charges, thereby allowing the particles to coalesce to larger particles. Upon coagulation, the rubber hydrocarbon is washed and processed further. To prevent coagulation when reacting with an acidic catalyst, a stabilizer in the form of a nonionic wetting agent must be added. Thereafter, the latex rubber is converted to an emulsion of cyclized rubber in a medium of 60 to 80 percent sulfuric acid. The reaction is sensitive to both the sulfuric acid concentration and the absolute temperature.

Equations (45) were developed for predicting the degree of conversion to cyclized units as a function of the reaction time and the aforementioned parameters.

In addition to the four basic methods of cyclization, there are other methods that are less commonly used, including cathode radiation (47). The use of high-tension alternating current in a hydrogen atmosphere with the rubber dissolved in Decalin solvent has also been reported (48). Decalin is a mixture of hydrogenated naphthalene derivatives.

Sensitizers for Negative Resists

The sensitizer or photoinitiator is the "heart" of the photoresist composition, if one may use this term. Resins with little or no intrinsic photosensitivity can be used in photoresist compositions and rendered insoluble in their ordinary solvents upon exposure with the proper sensitizer. Generally, sensitizers are employed in small quantities based on the weight of the resin; also they are proprietary and frequently quite expensive. They are present in all commercial photoresist compositions, even though the resin may have some intrinsic photosensitivity, to shorten exposure times and improve uniformity. Besides photosensitivity, the sensitizer compound must have certain physical properties. Before photoactivation, the sensitizer must be stable in the presence of the resin for long storage periods either in solution or in the dried film. When in solution and during application, the sensitizer should not tend to precipitate, crystallize, or otherwise concentrate itself. The resin and sensitizer must also be soluble in common solvents.

Sensitizers become activated by the absorption of light energy or of discrete wavelengths which with few exceptions lie in the near-ultraviolet region (2000 to 4000 Å). Once activated, the sensitizer transfers energy to the polymer molecule, which causes disassociation of the unsaturated group and subsequent insolubilization. Resin systems can be insolubilized by cross linking of linear molecules or by chain propagation and growth through activation of the unsaturated group.

The actual reactions that the sensitizers undergo are quite complex and the number of photosensitive compounds substantial. Sensitizers for photopolymerization that are activated from 2800 Å to within the visible range (4000 to 7000 Å) have been reported (49). Herein we will limit the discussion of sensitizers to known resin systems (Table 2-2) for brevity. Because of the complexity of photochemical reactions and the scope of the book, only simplified discussions of known reactions will be presented. Insolubilization of monomers is used primarily in solid- or dry-film resists, in whose section it will be discussed.

Some general terms can be used to express basic reactions that are known to occur. The initial step is the absorption of radiation of the proper wavelength to cause activation of the sensitizer. For sensitizer activation, it is essential that the light source used for exposure have a spectral output that at least overlaps the spectral response of the sensitizer. After absorption of the light energy, the sensitizer molecule is said to be in the "excited" state. In this state, no physical change has occurred, but the molecule is capable of undergoing reactions that would be unlikely to occur at its normal, or ground-level, energy state. Some of the reactions of the excited state are fruitful from the standpoint of insolubilization reactions, and some are not. The excitation energy of the sensitizer can be used to form reactive short-lived intermediate compounds with an odd number of electrons called "free radicals." Free radicals capable of insolubilization reactions are formed when the excited state is sufficiently long-lived and of sufficient energy. There are several different reactions by which free radicals can insolubilize resins (50), although not all substances that generate free radicals upon exposure make good sensitizers. Substances that produce comparatively stable radicals, such as triphenylmethyl, can actually retard polymerization (51).

Some reactions in the excited, or active, state do not involve free-radical formation in their energy tranfer. Absorbed energy can be reemitted at a longer wavelength or transmitted to other molecules in close proximity, which may not result in polymerization or cross linking. In spite of these negative-resist processes are very efficient, with numerous insolubilizing reactions occurring for each photon of light absorbed. Several thousand reactions per absorbed photon are not uncommon. Oxygen tends to interfere with some free-radical polymerization reactions and can affect exposure times. It is generally thought that the oxygen combines with radicals to form nonpropagating or chain-terminating peroxide compounds (52). Some unsaturated compounds are quite difficult to polymerize in air, but normally, unless projection printing is used, a gas impermeable layer in the form of the photomask is present during exposure. Patents have been issued (53,54) in which cover sheets are used to eliminate oxygen effects from photopolymerizable resins. Sensitizers with shorter-lived active states may be less sensitive to these reactions (55).

Sensitizers for Polyvinyl Cinnamate. There are a large number of compounds that reportedly sensitize polyvinyl cinnamate resins in the near-ultraviolet region and drastically increase the photosensitivity of the composition. On an exposure scale where dichromated shellac has a sensitivity rating of 30, unsensitized polyvinyl cinnamate has a rating of only 2.2. Many of the sensitizers increase the sensitivity of polyvinyl cinnamate to a rating of 300 or even a thousand or more. The basic types of sensitizers have been categorized in Table 2-4 along with references (37,56–58) to more definitive tables and patent literature.

Little specific information is available regarding the sensitizers actually used in the resist compositions. Levine, Lesoine, and Offenbach in their analysis of KPR-2 and KTFR (7) indicated that the former utilized a sulfur-bearing sensitizer containing about 10.0 wt percent sulfur. According to Robertson and West in Kodak's U.S. patent 2,732,301 (58), some compounds contain about this percentage of sulfur. Eisler indicated that the sensitizer in KPR might be crystal violet (59), a triphenylmethane dye (Table 2-4), patented by Minsk, Van Deusen, and Robertson (U.S. Patent 2,690,966). The dye is a relatively poor sensitizer unless aged with a peroxide compound, which drastically increases its photosensitivity in combination with polyvinyl cinnamate. Peroxides were reported in KPR (60), but there is no other evidence and no specific information on the Hunt Waycoat RC and 20 sensitizers.

The amount of sensitizer used with the polyvinyl cinnamate resist is based on the weight of the resin present. Some of the patents show as much as 10 percent of the weight of the polymer, but in actual practice, about 2 percent is used, based on published analyses. The real criterion would be based on the ability of the sensitizer to insolubilize the exposed areas of the resist film in a reasonable time, and it would appear that the selected sensitizers can perform this function when present in a concentration of only a few percent. It is probable that in some instances undesirable side effects may occur with an excess of sensitizer or that above a certain percentage the performance is not significantly improved. To sensitize partially esterified polyvinyl cinnamate resins, Murry and Leubner (6) indicated that thiazoline compounds are preferred.

Sensitizers for Diallyl Phthalates. Sensitizers listed for diallyl phthalate resins are basically the same as those listed for polyvinyl cinnamate resins. Azides, which were not listed for DCR (21,22), were patented by Kodak (29). A combination of sensitizers such as Michler's ketone, xanthone, and benzil were cited for use with diallyl isophthalate (23).

Sensitizers for Cyclized Rubber. Sensitizers for cyclized rubber have

TABLE 2-4 Sensitizer for Polyvinyl Cinnamate

Category	Reference	Example	Other compounds
Azido compounds	Hepher and Wagner British Patents 892,811 (37) and 892,812 (56)	4,4′-Diazidodibenzal acetone (speed: not listed)	British Patent 892,811—compounds containing two groups connected by a chain of three or more carbon atoms including a grouping —C—C═ British Patent 892,812—water-soluble forms
Nitro compounds	Minsk, Van Deusen, and Robertson U.S. Patent 2,610,120 (2)	*p*-Nitrodiphenyl (speed: 200)	Nitrobenzene—speed: 8 α-Nitronaphthalene—speed: 200 *p*-Nitrophenol—speed: 70 *p*-Dinitrobenzene—speed: 40
Nitro aniline derivatives	Minsk, Van Deusen, and Robertson U.S. Patent 2,610,120 (2)	4-Nitro-2-chloroaniline (speed: 300)	2,6-Dinitroaniline—speed: 70 5-Nitro-2-aminotoluene—speed: 200 2,4-Dinitro-6-chloroaniline—speed: 90 2,4-Dinitroaniline—speed: 90 2,4,6-Trinitroaniline—speed: 300
Anthrones	Minsk, Van Deusen, and Robertson U.S. Patent 2,670,285 (3)	2-Keto-3-methyl-1,3-diazabenzanthrone (speed: 500)	Dianthrone—speed: 80 Anthrone—speed: 10 1,9-Benzanthrone—speed: 100 2-Keto-3-azabenzanthrone—speed: 55 4-Cyclohexylbenzanthrone—speed: 3

Quinones	Minsk, Van Deusen, and Robertson U.S. Patent 2,670,286 (4)	9,10-Anthraquinone (speed: 100)	1,2-Benzanthraquinone—speed: 500 β-Chloroanthraquinone—speed: 200 β-Methylanthraquinone—speed: 5 1,4-Naphthoquinone—speed: 60 2,6-Dichloroquinone—speed: 9
Diphenyls	Minsk, Van Deusen, and Robertson U.S. Patent 2,670,287 (5)	Benzil (speed: 50)	Diphenyl ketone—speed: 20 4,4′-Tetramethyldiaminodiphenyl carbinol—speed: 100 4,4′-Tetramethyldiaminodiphenyl ketone—speed: 650 4,4′-Tetramethyldiaminodiphenyl methane—speed: 45 Diphenyl acetone—speed: 60 4,4′-Tetramethyldiaminobenzophenoneimide—speed: 100
Triphenyl-methane dyes	Minsk, Van Deusen, and Robertson U.S. Patent 2,690,966 (57)	Malachite green (speed initially: 10; after peroxide aging 40)	Peroxide aging: Before / After Victoria blue—speed: 5 / 70 Methyl violet—speed: 3 / 40 Crystal violet—XC speed: 15 / 250
Thiazoline compounds	Robertson and West U.S. Patent 2,732,301 (58)	1-Methyl-2-benzoyl-methylenenaphtho-thiazoline (speed: 1,400)	3-Methyl-2-propionylmethylenebenzothiazoline—speed: 140 2-Methyl-x-nitro-β-naphthothiazole—speed: 650 1-Ethyl-2-acetylmethylene-β-naphthiazoline—speed: 500

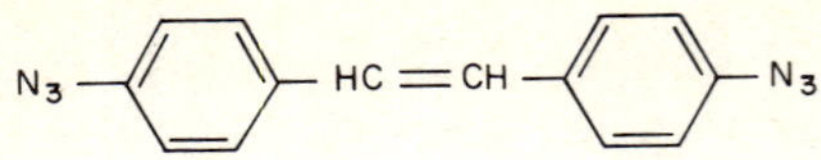

4,4′-Diazidostilbene (Reference 36)

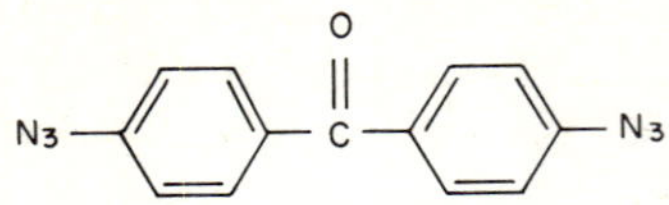

4,4′-Diazidobenzophenone (Reference 36)

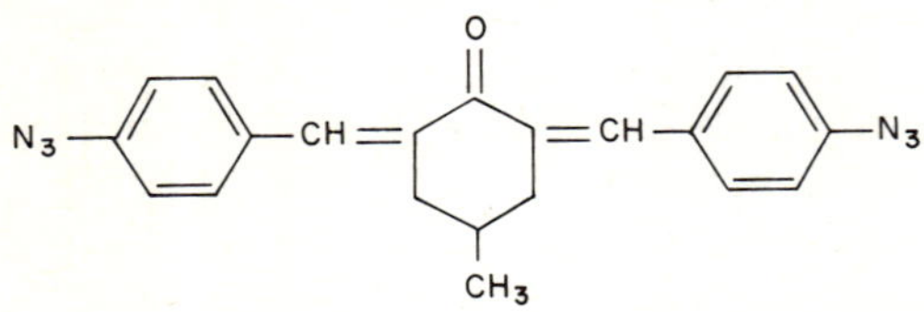

2,6-Di-(4′-azidobenzal)-4-methyl cyclohexanone (Reference 61)

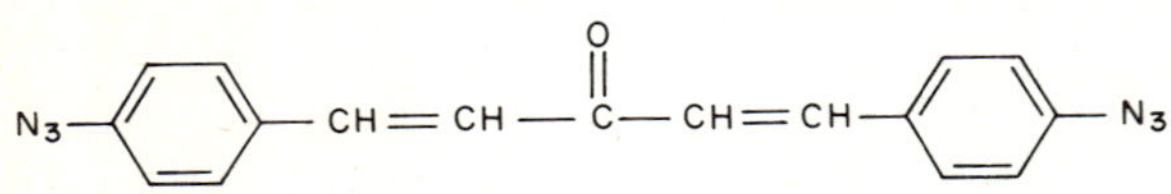

4,4′-Diazidodibenzalacetone (Reference 37)

Fig. 2-11 Structures of bis-azide sensitizers for cyclized-rubber resists.

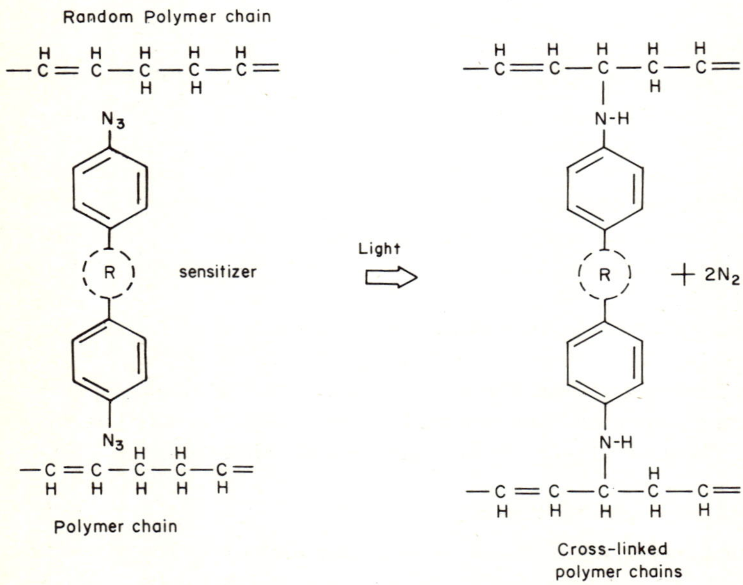

Fig. 2-12 A reaction of bis-azide cyclized-rubber sensitizers.

generally been limited to azides (36,61,62) (compounds containing one or more $-N_3$ groups). Some examples of compounds which have been used are shown in Fig. 2-11, and possible reactions in Fig. 2-12. Most of the compounds contain double bonds as well as azide linkages and are very active toward cross linking. Upon activation of the two double bonds and the two azide groups in the cyclohexanone sensitizer shown in Fig. 2-11, no less then six possible reaction sites are available for attachment and cross linking. Moreover, the sensitizer itself is a high-molecular-weight compound, and attachment to the polymer chain would assist in reducing water solubility. Sensitizers of this type can be used to render polymers containing no unsaturation, such as polyvinyl alcohol or gelatin, insoluble upon exposure to active light sources by cross-linking adjacent chains of the polymer through the formation of an amide (—NH— group). However, the sensitizer must be modified for use with a water-solvent system by the addition of polar groups such as sulfo or carboxylic groups. The 4,4′-diazidostilbene shown in Fig. 2-11 was used by Hepher (63) to prepare a water-base latex photoresist system by the substitution of $NaSO_3$-groups on the benzene rings. Today, similar compounds are used in the presensitized lithographic plate coatings and screen-stencil gelatins.

It may be noted that most of the compounds contain carbonyl groupings $(-\overset{\overset{\displaystyle O}{\|}}{C}-)$. This reportedly gives more efficient photo cross-linking activity (64). For example, the quantum yield of 4,4′-diazidostilbene is reported to be considerably lower than 4,4′-diazidobenzophenone. The only structural difference is the substitution of the carbonyl group for the ethylene grouping. Light sources producing radiation in the 2600 to 4700 Å range are capable of activating most of these compounds.

Of the group shown in Fig. 2-11, the cyclohexanone sensitizer was most suitable. It was reportedly the most soluble of those listed and resistant to the formation of crystals during storage in the liquid form and during coating operations. Poor sensitizer solubility limits the concentration of the polymer and promotes defects in the image by crystallization within the resist film.

Borden (65) described a nonazide sensitizer that could be used to render polyisoprene and natural rubber photosensitive. The method by which the sensitizer reacts was not stated, but it was reportedly capable of insolubilizing polymers not normally photosensitive, such as polyvinyl alcohol. Some examples of the sensitizers are shown in Fig. 2-13.

Additives for Negative Resists

As defined earlier, additives are substances added to the resist formulations to improve or enhance performance. One of the earlier problems encountered with polyvinyl cinnamate resists was a tendency not to develop

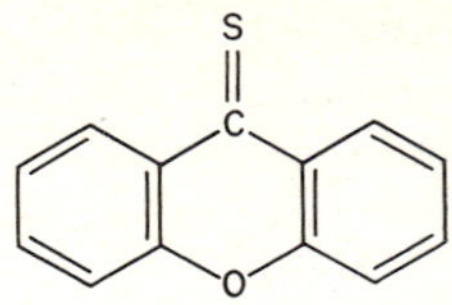

Thioxanthone

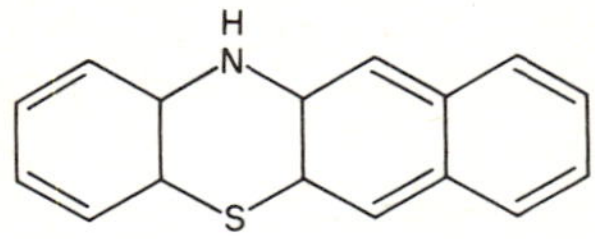

Benzo-(b)-phenothiazine

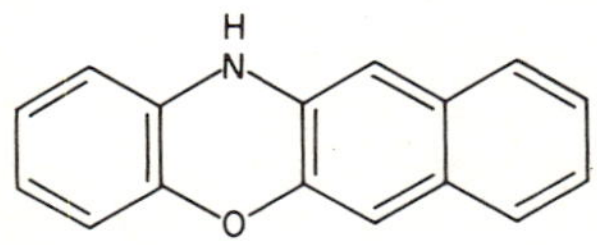

Benzo-(b)-phenoxazine

Fig. 2-13 Nonazide sensitizers from U.S. Patent 3,169,868 (65).

clearly if the coating had aged on lithographic plates for 24 h or longer. A residue in the form of a very thin scum remained in the unexposed areas; this caused etching and printing problems. The problem could be eliminated or greatly reduced by the addition of 0.5 to 1.5 g/l of triethanolamine (66) and a reducing agent, alone or in combination with organic acids (67), some of which are shown in Table 2-5.

A tendency of polyvinyl cinnamate coatings to darken after prolonged storage on copper plates was also a problem. Darkening was caused by dissolved copper in the coating and acted to reduce photosensitivity and to promote scumming and related problems. The situation could be alleviated by the addition of a poly- or dicarboxylic acid, such as oxalic, citric, or tartaric acids (68), particularly those containing hydroxy groups, such as the last two acids.

Thermal polymerization inhibitors which reduce the tendency of the resist coating to polymerize during preexposure processing were patented by Celeste (69). Examples cited included hydroquinone, pyridine, copper resinate, cuprous chloride, nitrobenzene, and β-naphthol.

The addition of a small amount of a phenol formaldehyde resin (novolak, see Fig. 2-20) to polyvinyl cinnamate resin photoresists was claimed to improve adhesion and reduce swelling during developing (70). A typical resist formulation is shown in Table 2-6 for the composition. The novolak resin was particularly helpful when the developer was trichlorothylene.

TABLE 2-5 Stabilizers to Reduce Scumming (67)

Stabilizers*	Acids
Phenol	Cinnamic acid
Hydroquinone	Benzoic acid
Catechol	Phenylacetic acid
Resorcinol	Hydrocinnamic acid
0-*tert*-Butyl-*p*-methoxy phenol	Chloroacetic acid
Phenyl-β-naphthylamine	Salicylic acid
p-Hydroxyphenylmorpholine	Crotonic acid
p-Phenylenediamine	

Combinations Cited (1 to 10 wt percent of polymer)
1. Hydroquinone
2. Four parts cinnamic acid One part hydroquinone
3. One part hydroquinone Ten parts benzoic acid Two parts cinnamic acid

* Stabilizers may be used alone or in combination with each other or the acids.

TABLE 2-6 Formulation Example of a Polyvinyl Cinnamate Resist Containing Novolak Additive (70)

Ingredient	Quantity
Polyvinyl cinnamate	208 g
Novolak (350–600 mol wt)	16 g
Photosensitizer (thiazoline type)	2.1 g
Hydroquinone	1 g
Monochlorobenzene	1,600 cm^3
Cyclohexanone	500 cm^3

Also, it could be added to the azide-sensitized rubber composition, which was similar to the KMER.

Solvents

Chemical solvents are used in the preparation of resins and constituents to hold them in solution during storage and application and for developing and related processing. In many instances, the solvents are the same for different resins; some of the more common are shown in Fig. 2-14. As before, solvents for the basic resin are considered individually. Further discussion on solvents with relation to their functions is contained in Chap. 4.

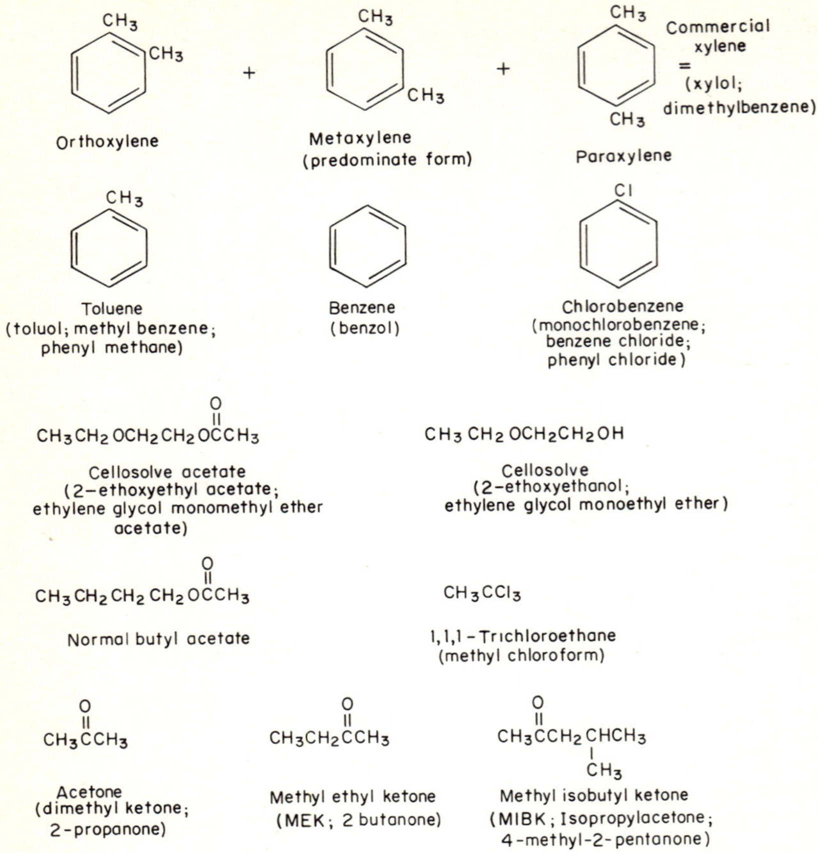

Fig. 2-14 Some common photoresist solvents.

Polyvinyl Cinnamate. A large number of solvents are available in patent literature for processing polyvinyl cinnamate photoresists, some of which are listed in Table 2-7 along with cited uses.

TABLE 2-7 Some Solvents Used with Polyvinyl Cinnamate

Solvent	Uses	U.S. Patent references
Pyridine	Reaction medium during esterification	2,610,120 (2)
Acetone	Dilution of pyridine in resin preparation	2,610,120 (2)
Three vol toluene One vol monochlorobenzene	Solvent mixture in which the polyvinyl cinnamate is dissolved	2,610,120 (2)

TABLE 2-7 (*Cont.*)

Solvent	Uses	U.S. Patent references
Nitrobenzene	Same as above and as a developer (2)	2,610,120 (2) 2,670,285 (3)
1,4-Dioxane	Solvent in which polyvinyl cinnamate is dissolved; solvent and developer for cellulose cinnamate	2,610,120 (2)
Acetic acid	Solvent and developer for polyvinyl cinnamate	2,610,120 (2)
Furfural	Solvent and developer for polyvinyl cinnamate	2,610,120 (2) 2,670,286 (4)
Cyclohexanone	Same as furfural	2,610,120 (2) 2,670,286 (4)
Benzaldehyde	Same as furfural	2,610,120 (2) 2,670,286 (4)
Methyl glycol acetate	Same as furfural	2,610,120 (2) 2,670,286 (3) 2,670,285 (4)
Morpholine	Same as furfural	2,670,286 (4)
Four vol xylene One vol methyl glycol acetate	Developing in tanks	2,670,286 (4)
Four vol xylene One vol methyl glycol acetate One vol isopropanol	Developing in tanks. Furfural or other lower alcohols may be substituted for alcohol.	2,670,286 (4)
Xylene Methyl hydrogen adipate	Developer	2,670,286 (4)
Xylene Isopropanol Phthalic acid	Developer. Other acidic compounds such as *p*-toluene sulfonic acid may be used. Cumine, ethyl benzene, butyl acetate, and others may be substituted for xylene.	2,670,286 (4)
Methanol	Added resist solvent system to increase the solubility of certain quinone- and nitro-group sensitizers	2,670,286 (4)

Phthalate Resins. Only diallyl phthalate and diallyl isophthalate will be considered in the prepolymer condition. Data listed are from Dynachem patents (21–23) and resin manufacturers' literature (24,25) in Table 2-8. In general, the diallyl phthalate resin is soluble in ester solvents, ketones, and aromatic hydrocarbons of low molecular weight; partially chlorinated hydrocarbon solvents; and the diallyl phthalate monomer, which is also a liquid. Partial solubility is reported (24) in toluene, xylene, and acetic acid.

Cyclized Rubber. As a hydrocarbon (a substance that contains only

TABLE 2-8 Some Solvents Used with Diallyl Phthalate Prepolymers

Solvent	Uses	References
Ethyl glycol acetate	Resin solvent	U.S. Patent 3,376,139 (21)
Methyl isobutyl ketone	Resin solvent and developing solvent	U.S. Patent 3,376,139 (21)
One vol methyl glycol acetate One vol toluene	Resin solvent	U.S. Patent 3,376,139 (21)
Xylene	Developing solvent	U.S. Patent 3,376,139 (21)
Two vol ethyl glycol acetate Three vol diacetone alcohol	Resin solvent	U.S. Patent 3,376,139 (21)
1,1,1-Trichloroethane	Developing solvent	U.S. Patent 3,376,139 (21)
Trichloroethylene	Stripping solvent	U.S. Patent 3,376,139 (21)
Isopropyl alcohol	Solvent for monomer and dilutant for other solvents	FMC (24,25)
Acetone	Resin solvent	(24,25)
Butyl and ethyl acetate	Resin solvent	(24,25)
Benzene	Resin solvent	(24,25)
1,4-Dioxane	Resin solvent	(24,25)

carbon and hydrogen), cyclized rubber is dissolved by hydrocarbon solvents. This is a large category of solvents that may be subdivided into aromatic and aliphatic solvents. Aromatic solvents are closed-chain-ring compounds, such as benzene, and the aliphatics are open-ring solvents, e.g., hexane or kerosine. Of the two groups, aromatic solvents are generally superior, having lower evaporation rates, and they have greater solvency power than aliphatics of the same boiling range. The solvent most often employed for cyclized-rubber resists is xylene. Commercial xylene, or xylol as it is sometimes called, is a mixture of the ortho- meta- and para-xylenes (Fig. 2-14), along with benzene and toluene impurities. Stoddard solvent is the most common aliphatic solvent used in processing cyclized-rubber resists. It is frequently used for developing when mixed with xylene in amounts from 30 to 50 vol percent. Stoddard solvent is a petroleum distillate that actually contains both aliphatic and aromatic compounds. A better grade solvent with a distillation range of 300 to 350°F may typically contain 45 percent aliphatics, 40 percent naphthalenes, and the rest other aromatics. An ASTM specification (D 484) has been prepared for the solvent. As indicated earlier, benzene is a solvent that can be used for cyclization.

Chlorinated hydrocarbons, such as monochlorobenzene, are also solvents for cyclized-rubber resists.

POSITIVE RESISTS

Positive resists are totally different from negative resists in response to actinic light and the resulting image, although the essential composition is similar; each contains sensitizers, resins, solvents, and additives. Positive resists are colored and soluble in strongly alkaline solutions. They develop in mildly alkaline solutions. General chemical resistance is less than the negative resists, and positives are more costly to procure. However, images from these resists are extremely accurate, require minimal processing technique, and involve few processing steps. Moreover, the use of aqueous stripping and developing solutions greatly simplifies the equipment selection by allowing low-cost readily available plastics to be used. Problems from the use of flammable solvents are minimized with positive resists, and conveyerized spraying can be utilized for both stripping and developing.

It was stated earlier that the essential composition of positive resists is similar to the negative resists in that they both contain sensitizers, resins, and additives dissolved in a solvent system. It is apparent, however, that the nature and functions of at least part of these constituents differ substantially from their counterparts in the negative resists. If polymerization occurs in the exposed areas of the negative resists, one might naturally assume that depolymerization was occurring with positive resists. In other words, the assumption would be that the sensitizer reacts with absorbed radiation to reduce the molecular weight of the resin. Such is not the case, however.

Sensitizers in positive resists do react with the absorbed radiation, although not to depolymerize the resin but rather to change the solubility of the coating from solvent-soluble to water-soluble. Resins used in positive resists contain no unsaturation and, in fact, do not react with the sensitizer during exposure. Low-molecular-weight resins are generally employed to ensure good aqueous developing characteristics. Both the resin and the sensitizer are good film-forming agents with a much higher ratio of sensitizer on a weight basis than the negative resist. For example, most negative resists contain 50 to 70 times as much resin as sensitizer, whereas the positives may only contain 2 to 4 times as much resin as sensitizer. This is necessary because the sensitizer does not react with the resin during exposure and must therefore bring about the change in coating solubility characteristics by itself. From its functions, it is apparent that the sensitizer has resinlike properties and is the primary constituent of the reist. As such, it will be discussed first.

Information regarding resist composition is less than that for negative resists although positive images have been produced on paper since the 1920s, starting with the Ozalid process (71). Resist coatings were originated for positive lithographic plates, as were the negative resists, and they have been available for many years. It was not, however, until the early 1960s that positive photoresists were introduced to the electronic industry by the Shipley Company through a licensing agreement with Azoplate Corporation. For several years the Shipley AZ resist series was the only available positive resist. Several improvements were made by the Shipley Company with the cooperation of Azoplate Corporation during this period. In the latter part of the 1960s, Eastman Kodak, GAF, P. A. Hunt, and Dynachem Corporation introduced positive resists, also. There is no doubt that the introduction of positive resists was a significant contribution to the involved industries and a major milestone in the evolution of photoresists.

Sensitizers for Positive Resists

Most sensitizers are derivatives of compounds variously called "diazo oxides" or "orthoquinone diazides" (72). They can be prepared in both naphthalene and benzene forms (Fig. 2-15), and one or more of these are usually joined to a central structure (Fig. 2-16) through an ester or amide linkage. The compounds are usually yellow in color, stable and resistant to decomposition, soluble in solvents of low polarity, and insoluble in water. Naphthalene derivatives are apparently preferred for photoresists (73), and few specific examples of sensitizers are described in literature other than patent literature.

There are described, however, many types of diazo compounds other than the diazo oxides or quinone diazide compounds. Saunders (74) indicated that sufficient information was available on diazo compounds to fill some 15,000 pages. That statement was referring to information available *before* 1936. Considering the enormous information generated since then, it is likely that several times that total is available now. It is therefore necessary to limit the discussion to the most salient compounds.

The quinone diazides shown in Fig. 2-15 are all possible selections as the light-sensitive elements joined to central groupings, examples of which are shown in Fig. 2-16. It should be noted that the light-sensitive elements are joined in every instance through an ester or an amide linkage to the central group. This is accomplished by reacting the sulfochloride ($-SO_2$

Napthoquinone diazides

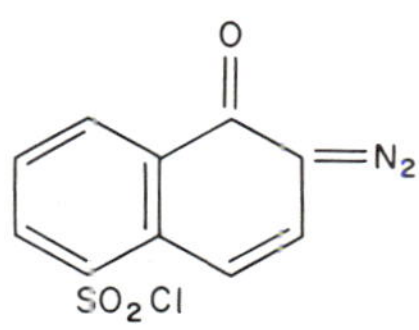

Napthoquinone-1,2-diazide-5-sulfochloride

Benzoquinone diazides

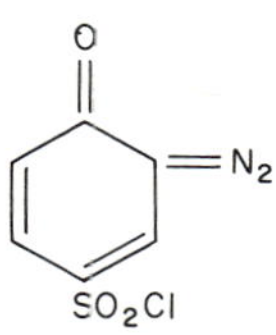

Benzoquinone 1,2-diazide-4-sulfochloride

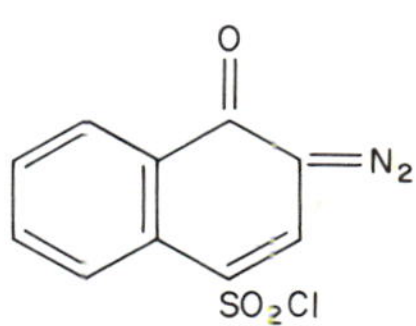

Napthoquinone-1,2-diazide-4-sulfochloride

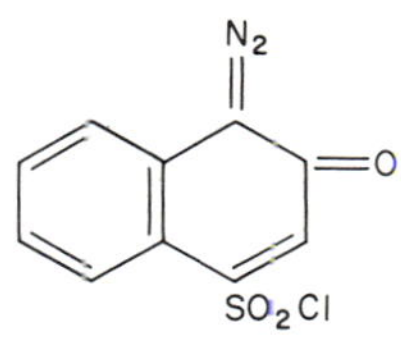

Napthoquinone-2,1-diazide-4-sulfochloride

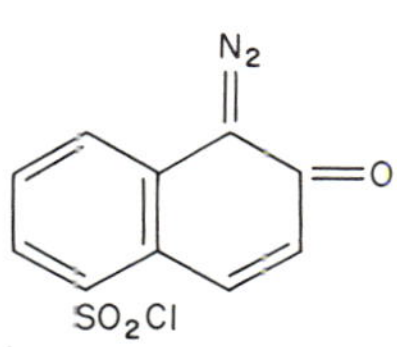

Napthoquinone-2,1-diazide-5-sulfochloride

Fig. 2-15 Positive-resist light-sensitive elements.

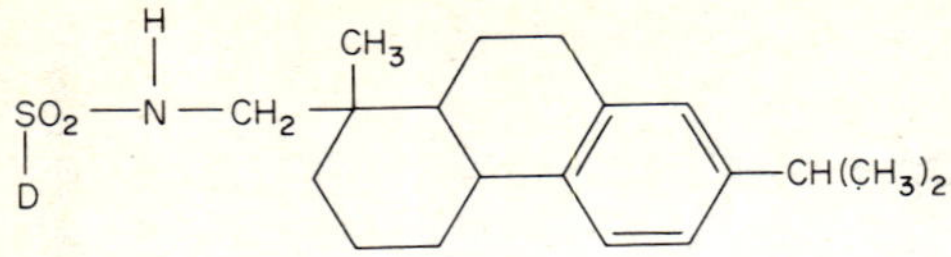

U.S. Patent 2,797,213 (1957).
Reference 76. (GAF Corp.)

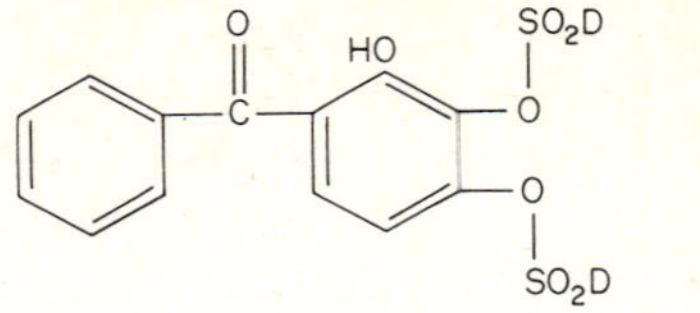

U.S. Patent 3,148,983 (1964).
Reference 79. (Azoplate Corp.)

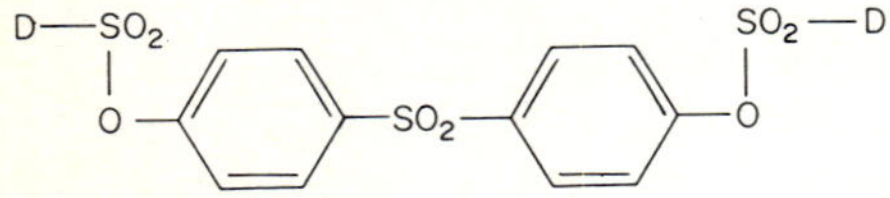

U.S. Patent 3,046,118 (1962).
Reference 80. (Azoplate Corp.)

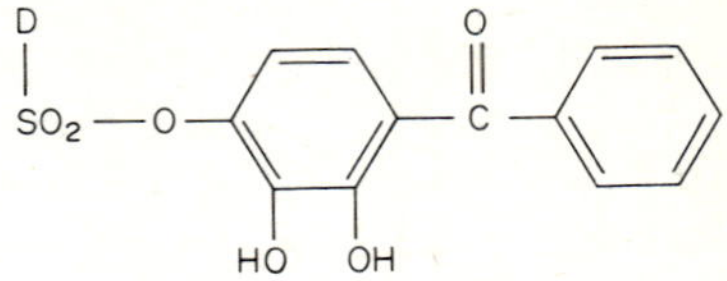

U.S. Patent 3,106,465 (1963).
Reference 78. (Azoplate Corp.) For metal-etching resist composition, see U.S. Patent 3,201,239 (1965).
Reference 95. (Azoplate Corp.)

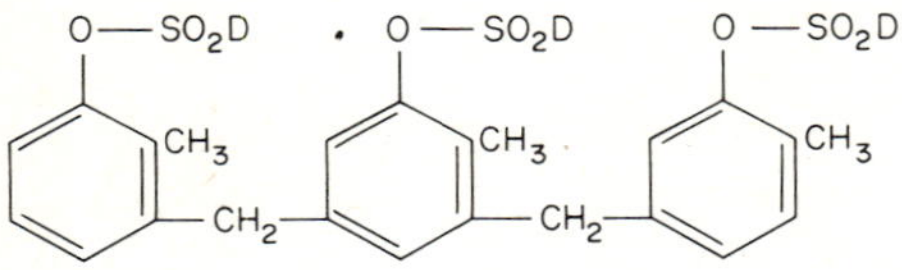

U.S. Patent 3,046,120 (1962). Reference 85. (Azoplate Corp.) Single-component resist made by direct esterification of the hydroxyl groups of O-cresol novolak with a slight excess of naphthoquinone diazide sulfuryl chloride.

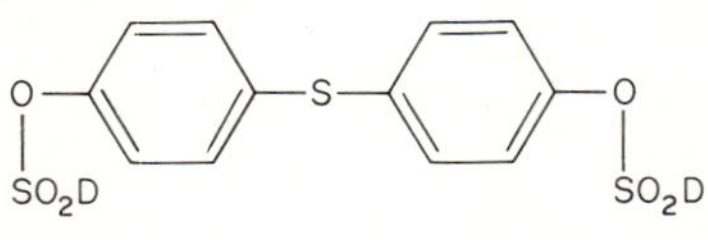

U.S. Patent 3,184,310 (1965).
Reference 81. (Azoplate Corp.)

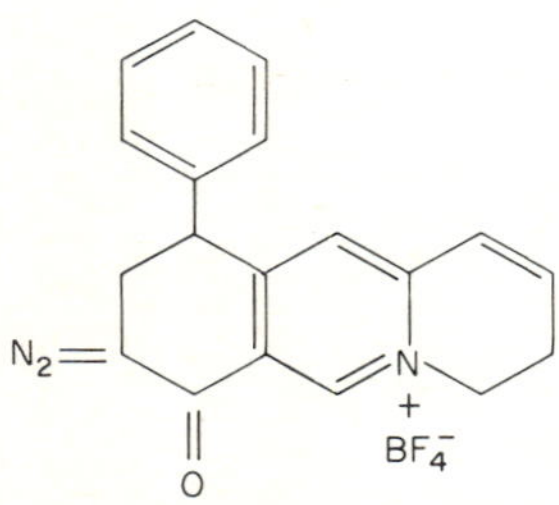

South African Patent 68 01 224.
Reference 83. (Kodak) Also see U.S. Patent 3,567,453 (1970).

Fig. 2-16 Central groups of sensitizers with naphthoquinone diazide light-sensitive elements (D).

Cl) on the light-sensitive element with a hydroxy (–OH) or amine (—NH_2) on the central group. Elements with carboxy acid chloride derivatives can also be prepared. Figure 2-17 shows the preparation of naphthoquinone diazide sulfonyl chloride. Hence, the elements are shown as their

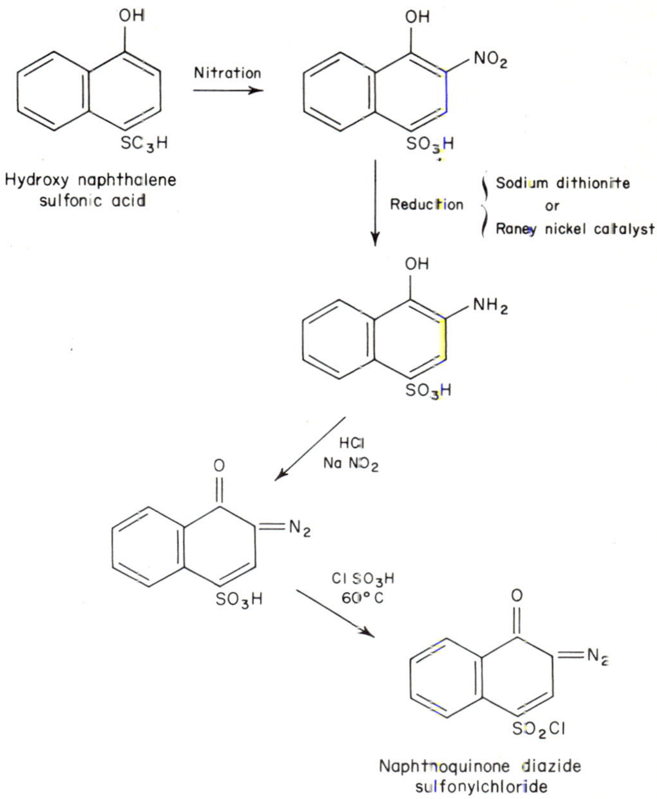

Fig. 2-17 Preparation of light-sensitive elements for positive resists.

chloride derivatives. Of the four naphthoquinone isomers (compounds which have the same formula but different structure and properties), the one listed first is the most widely used, judging by the patent literature (73,75–83), followed by second of the series. Little use is apparently made of the last two members of the series, although all are photosensitive. Benzoquinone diazides are also used (84–86).

Upon reacting with light, the quinone diazides undergo a drastic physical change; this involves rearrangement of the ring structure containing the nitrogen and oxygen to form a new compound (87) with the evolution of

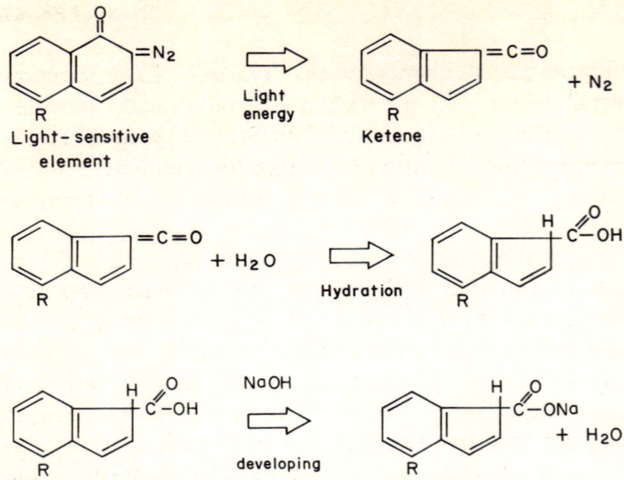

Fig. 2-18a Reaction sequence proposed by Sus in 1944. The ketene formed is very highly reactive and has not been detected. It would appear that the moisture would have to be present in the film and react immediately.

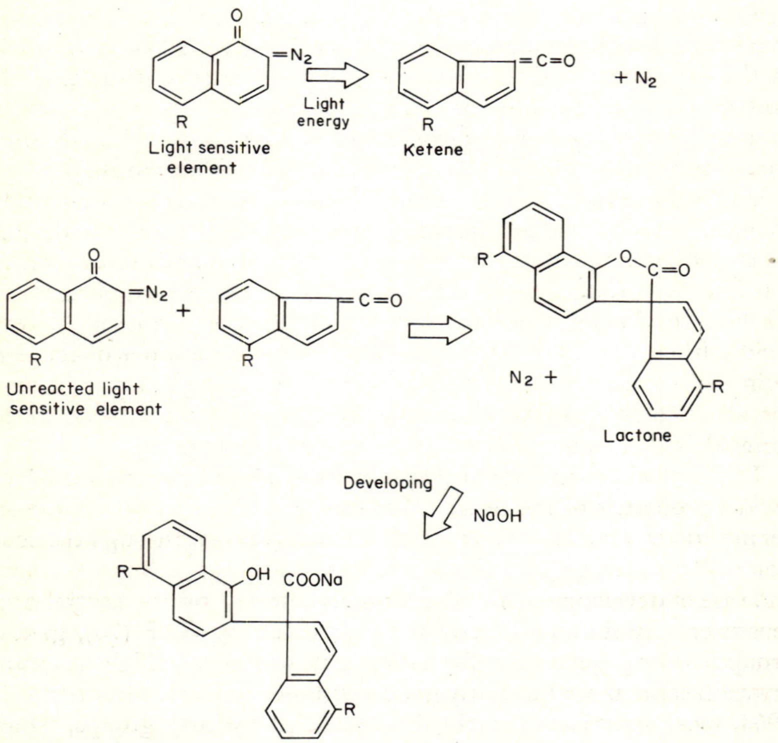

Fig. 2-18b Reactions suggested by Levine for the exposure of positive resists in 1969, as interpreted by the author. Initially, the surface layer reacts to form the ketene. This in turn reacts with unexposed sensitizer molecules to form a lactone. During developing, the lactone ring is broken, and a carboxylic acid group salt is formed.

nitrogen. Figure 2-18 illustrates the reaction for the compounds of interest herein.

In Fig. 2-18*a*, the process described by Sus (87) is described. This was generally accepted until Levine (96) suggested that a lactone rather than a carboxylic acid group could be formed during the exposure reaction, as shown in Fig. 2-18*b*. Lactones are ring compounds frequently made from short-chain aliphatic hydroxy acids having hydroxyl groups at the end opposite a carboxylic acid group. An internal ester is formed by heating with the evolution of water. Strong caustics reverse the reaction by breaking the ring and forming the hydroxyl group, again with a carboxylic acid salt. Both concepts start with the formation of a highly reactive compound called a "ketene," upon exposure of the naphthoquinone diazide light-sensitive element. [Ketenes are compounds containing a carbonyl group attached to another carbon atom by a double bond (R=C=O)]. Ketenes apparently react so rapidly that they cannot be detected during exposure. Sus indicated that the ketene reacts with water to form a carboxylic acid group. The acid is then neutralized by a caustic developing solution. Levine also suggested that the ketene is formed but that it reacts with an unreacted naphthoquinone diazide sensitizer element as the exposure reactions proceed downward from the surface. It was pointed out that this type of compound would be more resistant to dissolution by caustics than carboxylic acid compounds because salt formation would occur more readily than the rupture of the lactone ring.

Unlike the negative resists, the image (unexposed) area can also undergo a chemical reaction called "coupling," in which the diazide joins with certain other compounds notably phenols and aromatic amines in the form of an azo linkage. Coupling, illustrated in Fig. 2-19, occurs instantly in the alkaline developing solutions, which also removes the exposed residues. Acid solutions do not promote coupling, and acid solutions of couplers and quinone diazides are stable in this respect. Coupling reduces the tendency for alkali-soluble resins containing phenols or amines, such as phenol formaldehyde resins, to leach out during developing.

The central group, which may not itself be photosensitive, affects the overall properties of the resist. For example, the solubility of the sensitizer in various solvents can be controlled to some extent through the composition of the central group. Film-forming ability, crystallization tendencies, and ease of development are also strongly affected by the central group of sensitizer. Rosin-amide derivatives, patented by GAF (76), give center groups having good film-formation properties and high resistance to crystallization as well as fairly good resistance to caustic solutions. After 1963, most sensitizers contained unesterified hydroxy groups, which improved the developing characteristics of the exposed films. There was also an incompatibility with some of the common phenol formaldehyde

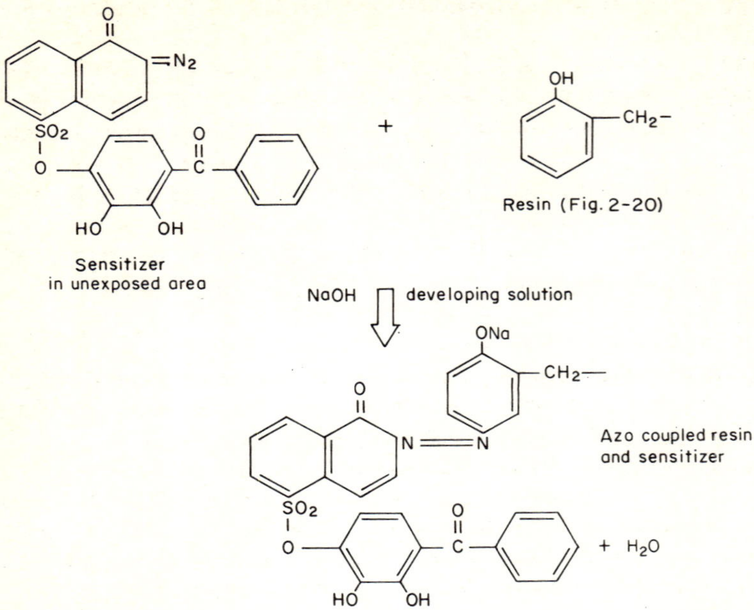

Fig. 2-19 Reaction of sensitizer molecule with the resin in the unexposed areas.

resins that are often added (see discussion of resins) to improve adhesion and film-forming properties. Resins tend to cause precipitation of some sensitizers from solution and sensitizer crystallization in dry films. Improvement was noted when the sensitizer contained a carbonyl $(R-\overset{\overset{O}{\|}}{C}-R')$ group in close proximity to the hydroxy group (Fig. 2-16, "ketones"), and greater amounts of resin could reportedly be used (80). Sulfur-containing sensitizers (81) were noted for their stability and clean developing. Sulfur compounds, such as thiourea, have long been used as stabilizers for diazo compounds. Mixtures of compatible sensitizers have been proposed (83).

Diazide-type sensitizers generally decompose upon heating rather than melting. Lower-molecular-weight sensitizers start to decompose at about 150°C, whereas others are stable above 300°C for short periods. Safe baking temperatures would appear to be 120°C or preferably less (100°C) for the sensitizers. Elevated baking temperatures may also tend to harden some of the resins to prevent clean developing and give the impression of lost photosensitivity if done before exposure.

Information has not been released regarding the composition of Kodak positive resist. A combination of quinone diazide derivatives of anthracene and a phenol formaldehyde resin was patented (83) by a group from Kodak for use as a positive-working photoresist. Sensitizer examples are shown in Fig. 2-16. Azide derivatives in combination with polyamide resins were also patented (88) as a positive-resist composition. Kalle Co. AG also patented (89) a positive-resist composition using azides. Information regarding the Dynachem positive resist is also unavailable. The sensitizer appears to be a chainlike structure either made up of repeating photosensitive elements or with a vinyl chain to which light-sensitive elements are attached.

Resins for Positive Resists

In positive resists, resins serve to improve adhesion of the coating composition, fortify the coating for better chemical resistance, reduce the tendency for the sensitizer to crystallize out of the coating, increase the viscosity and coating characteristics, and reduce the cost of the resist solution as a partial substitute for the resinlike sensitizer. Resins in positive resists, unlike those in negative resists, are not photosensitive as a rule but have certain other characteristics. In general, the resins are not extremely hydrophobic and are soluble in alkaline solutions but are resistant to acids. To be extremely hydrophobic would severely limit the amount that could be used and still obtain clean developing. If, however, the resins are too soluble in alkaline solutions, they could be leached from the coating in the unexposed areas; no useful image would be obtained. The resin combinations must therefore be as hydrophobic as possible but still developable in alkaline solutions. Resins should contain no chemical unsaturation (83).

The sensitizers are diazo compounds which are capable of forming an azo dye with certain other compounds through the coupling reaction mentioned earlier and illustrated in Fig. 2-19. Resins containing phenols (Fig. 2-20) are especially useful in coupling reactions. Aromatic amines ($R—NH_2$) are also capable of undergoing coupling reactions, as are ethers formed from phenols.

By far, the resin most often mentioned for use with positive-resist sensitizers is the reddish-colored phenol formaldehyde novolak and related derivatives. The resins are adherent to most metal surfaces and are good film-forming agents. Moreover, they are soluble in alkaline solutions and many common solvents, and they are capable of coupling with the sensitizer (Fig. 2-19). Novolaks are linear and permanently fusible and do not form cross-linked polymers. Metacresol, a derivative of phenol that is more reactive, is often used to prepare novolak resins, as shown in Fig. 2-20. When a methyl group is located at the meta (3 or 5) position, the reactivity of the compound toward polymerization with formaldehyde is greatly

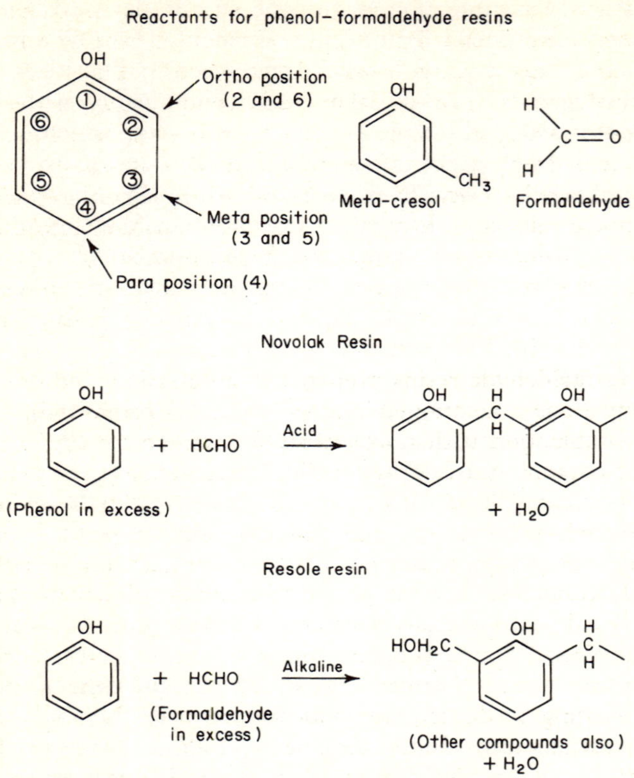

Fig. 2-20 Phenolformaldehyde resins.

increased. Resins made with metacresol are also more moisture-resistant that those with phenol. Novolak resins have long been known and prepared commercially (90) as a shellac substitute, although they were not particularly successful as such. The ability to resist further polymerization is attributed to the fact that the chains terminate with phenol groupings, having been prepared with an excess of phenol (Fig. 2-20). Preparation takes place in an acid solution, but it is possible that when added to resists the hydrogen atoms on some of the hydroxyl groups are replaced by a sodium ion. This would tend to give more rapid dissolution and coupling in developing solutions.

If the reacting medium of phenol formaldehyde mixtures is alkaline in nature and an excess of formaldehyde is used, the resin formed is capable of undergoing further polymerization upon heating. This resin, termed a "resole," terminates in a reactive alcohol group rather than in a phenol (Fig. 2-20). When heated, two alcohol groups combine and evolve a molecule of water to form an ether ($R-O-R_2$) linkage and thereby bring about an increase in molecular weight. Continued heating causes further polymerization into a hard cross-linked structure that is insoluble in caustics or solvents. In the initial or resole condition, the material may be either liquid or solid, is soluble in solvents and alkaline solutions, and is said to be an A-stage resin. The second, or B, stage is a partially polymerized condition in which the resin is still soluble in organic solvents but generally not in alkaline solutions. Thermal transition into this stage is usually accompanied by some flowing or apparent melting. B-stage resins are used as adhesives when impregnated on various fiber materials. Upon further heating, the final, or C-stage, resin called a "resit" is formed. In this condition, it is resistant to acids, alkalines, and solvents. The use of phenol formaldehyde resins prepared in an alkaline medium in photoresist formulations is mentioned in a Kalle Co. AG patent (86). Polyvinyl ethers in combination with novolaks were patented by Christensen (91). Steinhoff, Isaacson, and Roelants of the Shipley Company mention vinyl ethers in a patent (94) on roller coating. Lower alkly polyvinyl ethers, such as methyl, ethyl, butyl, and isobutyl, were specifically mentioned (Fig. 2-21). Polyvinyl ethers are used in formulations of adhesives to impart tack (stickiness) and for plasticizing action. They are known (92) to be compatible with phenol resins and are stable and insoluble in highly alkaline solutions, though they are less resistant to acids. Additions to novolak resins would therefore be beneficial from the standpoint of improved coating flexibility and adhesion to metal surfaces as well as improved resistance to mildly alkaline solutions. Alkali-soluble resins (e.g., novolak) and the sensitizer which is alkali-soluble after exposure can be used in combination with some resins which are not soluble in alkaline solutions provided the ratio is not too great (73). Amounts up to the weight of the alkali-soluble constituents apparently can be used, but scumming in the form of a thin undissolved film may occur. If the amount of non-alkali-soluble resin (e.g., polystyrene) is less than half of the weight of the alkali-soluble constituents, the problem is alleviated. These resins must also be good film formers and resistant to further polymerization. The advantage of these resin additions is primarily increased coating thickness, but chemical resistance will also be improved.

Styrene, methyl styrene, and styrene–maleic anhydride copolymers in combination with novolaks were mentioned in several patents (73,81,91) of both Shipley and Kalle Co. AG. Styrene resins (polyvinyl benzene) are hydrophobic but soluble in a variety of organic solvents, and they can be emulsified in aqueous alkaline solutions (93) with soaps and wetting agents. Where copolymerized with maleic anhydride (Fig. 2-21) a resin that is readily soluble in alkaline solutions is obtained. Carboxylic acid groups form water-soluble sodium salts that allow ready dissolution of the resin.

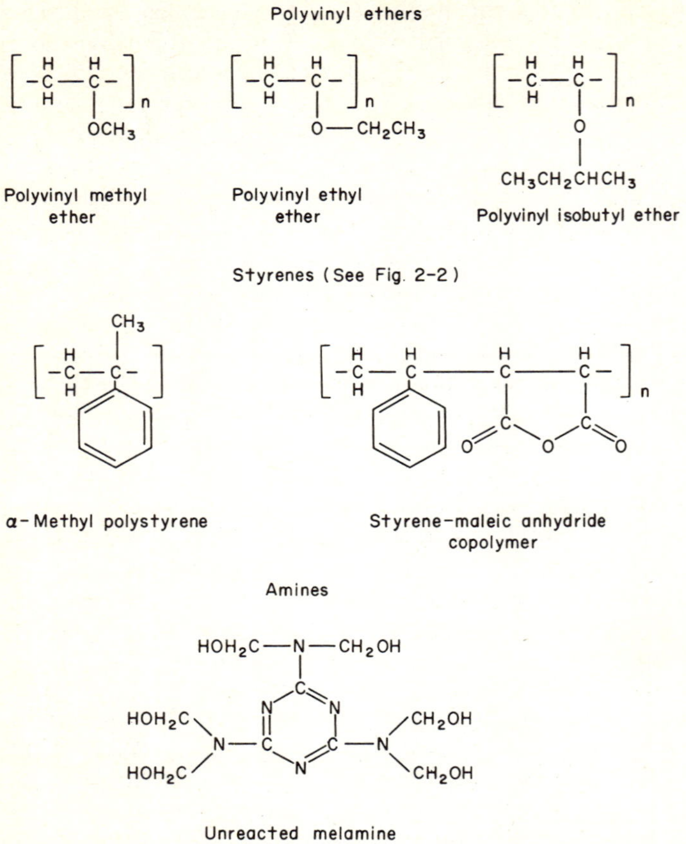

Fig. 2-21 Supplementary resins reportedly usable with positive resists.

The copolymer resin is clear, brittle, hard, and soluble in some organic solvents. Styrene-acrylic or -methacrylic acid copolymers can also be used (97).

Melamine resins, more specifically a melamine resin modified with butyl alcohol, were also shown to be compatible resins for positive-resist compositions. In addition to being combined with novolaks alone, a tri-resin group with sucrose benzoate could also be used.

Unalkylated melamine resins cure by heat alone to an insoluble, unfusible solid. The presence of alkyl groups tends to retard this effect, and the addition of acids is required for thorough curing. Melamine resins have high heat and water resistance and form tough films adherent to metals. Alkylated melamine coatings are fairly common additives to high-temperature-baking alkyd enamels, especially the butylated resins. Melamine alkyd coatings are used as automotive finishes. Alkylated melamine resins are soluble in xylene among other solvents.

Solvents for Positive Resists

Solvents for positive resists must be selected carefully. Unlike the negative resists, large amounts of sensitizer are present and may tend to crystallize from the solution either from incompatibility with the type or amount of resin present or because of the relatively large quantity maintained in solution. The amount of resin and the type are predicated upon the quantity of sensitizer that can be safely maintained in solution. Two or three percent solutions, which is generally too low to be used in circuit-board work, are often mentioned in patent literature.

As indicated in the section on negative-resist solvents, the amount and type of solvents play an important role in the formation of a uniform coating, and it is probable that a combination of solvents is present. The true solvents are the least volatile, to allow the coating to level itself before evaporation solidifies the film, which is invariably uneven immediately after the application process. Lower-boiling solvents may tend to allow drying of the film prematurely. Positive resists, however, are quite resistant to the formation of a surface skim during drying.

From the patent literature, the most prevalent solvent used to dissolve the resinlike sensitizers is ethylene glycol monomethyl or monoethyl ether (Cellosolve or methyl Cellosolve). For more hydrophobic resin additives, Cellosolve acetate (ethylene glycol monoethyl ether acetate), which is commonly used as a true solvent for negative-resist resins, is employed. Ester solvents, particularly acetates such as butyl acetate, are mentioned as dilutants, as are aromatic hydrocarbons, such as xylene (see Fig. 2-14).

Mentioned prominently during preparation as a solvent was dioxane.

Additives for Positive Resists

One of the more prevalent problems with positive resists has been their instability, and several stabilizing additives for diazo compounds have been mentioned. Dyes and resin plasticizers may also be present.

Stabilizers are usually reducing agents or antioxidants, such as thiourea (77), sugar, and related compounds. Organic acids (77,91) were also mentioned as stabilizers. Blue or red oil-soluble dyes along with their solubilizing oils (73) have also been mentioned as possible additives. Because dyes may tend to alter the exposure characteristics, they are used sparingly. Water-soluble dyes that could be leached from the coating have been avoided. An azo dye of sorts is formed during development of the unexposed areas of positive reists by the coupling reaction (Fig. 2-19) between the diazo sensitizer and the novolak resin. Further, both the sensitizer (quinone diazides) and the basic resin (novolak) are also colored.

Sensitizer molecules would in general be strongly yellow in color with the novolak having a reddish color, which would predominate in the mixture. Examples (79) of oil-soluble dye that can be used with positive-resist compositions are methyl violet BB in castor oil and roseaniline hydrochloride in sesame oil. About $3\frac{1}{2}$ wt percent dye of the sensitizer-resin combination was mentioned.

Resin plasticizers were disclosed by the Shipley Company (73). Phthalate esters were mentioned specifically, although novolaks can be hardened by the use of hexamethylenetetramine. The addition of a small amount of resole resin to the novolaks could function as a cross-linking agent to give a harder coating with good chemical resistance.

REFERENCES

1. L. M. Minsk, U.S. Patent 2,725,372 (1955).
2. L. M. Minsk, W. P. van Deusen, and E. M. Robertson, U.S. Patent 2,610,120 (1952).
3. L. M. Minsk, W. P. van Deusen, and E. M. Robertson, U.S. Patent 2,670,285 (1954).
4. L. M. Minsk, W. P. van Deusen, and E. M. Robertson, U.S. Patent 2,670,286 (1954).
5. L. M. Minsk, W. P. van Deusen, and E. M. Robertson, U.S. Patent 2,670,287 (1954).
6. J. J. Murry and G. W. Leubner, U.S. Patent 2,739,892 (1956).
7. H. A. Levine, L. G. Lesoine, and J. A. Offenbach, *Kodak Photoresist Semin. Proc.*, vol. I, pp. 15–17, 1968, *Kodak Publ.*, p. 192A.
8. G. M. J. Schmidt, *J. Chem. Soc.*, p. 2014, 1964.
9. F. I. Sonntag and R. Srinivasan, *Tech. Pap., First Reg. Tech. Conf. Soc. Plast. Eng., Mid-Hudson Sec.*, November, 1967.
10. S. Kikuchi and K. Nakamura, *ibid.*
11. A. Murry, U.S. Patent 1,965,710 (1934).
12. C. F. N. Allen and J. A. van Allen, U.S. Patent 2,566,302 (1951).
13. C. C. Unruh and C. F. N. Allen, U.S. Patent 2,716,097 (1955).
14. C. C. Unruh and C. F. N. Allen, U.S. Patent 2,716,102 (1955).
15. C. C. Unruh and C. F. N. Allen, U.S. Patent 2,716,103 (1955).
16. C. C. Unruh and C. F. N. Allen, U.S. Patent 2,706,725 (1955).
17. T. Tsunoda and S. Nozoki, *Bull. Tech. Assoc. Graphic Arts Japan*, vol. 5, pp. 1–4, 1962.
18. R. F. Kelly, *Proc. Second Kodak Semin. Microminaturization, Kodak Publ. P*-89, 1966, p. 31.
19. R. H. Collins and F. T. Deverse, U.S. Patent 3,549,368 (1970).
20. I. W. Martinson, "Photoprinting and Photoresists," *Spec. Tech. Semin. Inst. Printed Circuits*, Sept. 16–19, 1968, Los Angeles, Calif. (Proceedings, published January, 1969, sec. B, contains only excerpts of the talk, and the "diallylphthalate" statement is not present.) Also see J. F. Herzog, *EIPC Conf. Printed Circuit Manufacturing*, Feb. 19–20, 1970, pp. 108–115.
21. M. N. Gilano, I. W. Martinson, and L. H. Ott, U.S. Patent 3,376,139 (1968).
22. M. N. Gilano and I. W. Martinson, U.S. Patent 3,376,138 (1968).
23. M. N. Gilano, I. W. Martinson, and L. H. Ott, U.S. Patent 3,462,267 (1969).
24. FMC bulletin on DAPON (diallyl phthalate), Food Machinery Corporation.
25. FMC bulletin (*Pamphlet* 28) on diallyl isophthalate, Food Machinery Corporation.
26. C. A. Heiberger, U.S. Patent 3,096,310 (1963).
27. P. E. Willard, U.S. Patent 3,030,341 (1962).

28. H. Strange, W. B. Tuemmler, and J. F. Allen, German Patent 1,141,276 (1962).
29. F. J. Rauner, U.S. Patent 3,476,176 (1969).
30. E. M. Robertson and J. A. van Allen, U.S. Patent 2,870,011 (1959).
31. S. H. Merrill and D. A. Smith, U.S. Patent 2,861,057 (1956).
32. W. Schellenberg and O. Bayer, U.S. Patent 3,030,208 (1962).
33. M. C. Agens, U.S. Patent 3,031,301 (1959).
34. W. Kornfeld, *Proc. Second Kodak Semin. Microminiaturization, Kodak Publ. P*-89, 1966, pp. 23–30.
35. D. H. Gittleman, *Circuits Manufacturing*, vol. 10, no. 10, pp. 52–57, 1970.
36. M. Hepher and H. M. Wagner, U.S. Patent 2,852,379 (1958).
37. M. Hepher and H. M. Wagner, British Patent 892,811 (1962).
38. J. D. D'Ianni, F. J. Naples, J. W. Marsh, and J. L. Zarney, *Ind. Eng. Chem.*, vol. 38, pp. 1171–1181, 1946.
39. R. K. Agnihotri, D. L. Falcon, F. P. Hood, L. G. Lesoine, C. D. Needham, and J. A. Offenbach, "Structure and Behavior of Cyclized Rubber Photoresist," *Tech. Pap., Reg. Tech. Conf. Photopolymers, Mid-Hudson Sec. Soc. Plast. Eng.*, 1970, pp. 33–39.
40. P. J. Flory, *J. Am. Chem. Soc.*, vol. 61, p. 1518, 1939.
41. W. J. S. Nauton, *The Applied Science of Rubber*, Edward Arnold (Publishers) Ltd., London, 1961, pp. 90–98.
42. H. A. Bruson, L. B. Sebrell, and W. C. Calvert, *Ind. Eng. Chem.*, vol. 19, p. 1033, 1927.
43. H. L. Fisher, *Ind. Eng. Chem.*, vol. 19, p. 1325, 1927.
44. H. J. J. Janssen, *Rubber Age*, vol. 78, pp. 718–722, 1956.
45. M. Gordon, *Ind. Eng. Chem.*, vol. 43, pp. 386–393, 1951.
46. M. Gordon, British Patent 634,879 (1950).
47. E. B. Newton, U.S. Patent 1,906,402 (1933).
48. H. L. Fisher, *Chemistry of Natural and Synthetic Rubbers*, Reinhold Publishing Corporation, New York, 1957, p. 175.
49. A. B. Cohen, *SPSE Second Symposium on Unconventional Photographic Systems*, 1967, pp. 122,123.
50. D. W. Woodward, V. C. Chambers, and A. B. Cohen, *PS & E*, vol. 7, pp. 360–368, 1963.
51. G. Oster and Nan-Loh Yang, *Chem. Rev.*, vol. 68, no. 2, pp. 125–151, March, 1968.
52. N. G. Gaylord and H. F. Marks, *Linear and Stereoregular Addition Polymers*, Interscience Publishers, Inc., New York., 1959, pp. 8–10.
53. R. B. Heiart, U.S. Patent 3,060,026 (1962).
54. M. Burg, U.S. Patent 3,203,805 (1965).
55. R. O. Kan, *Organic Photochemistry*, McGraw-Hill Book Company, New York, 1966, pp. 78–80.
56. M. Hepher, British Patent 892,812 (1962).
57. L. M. Minsk and W. P. van Deusen, U.S. Patent 2,690,966 (1954).
58. E. M. Robertson and W. West, U.S. Patent 2,732,301 (1956).
59. P. Eisler, *The Technology of Printed Circuits*, Heywood and Co., Ltd., London, 1959, p. 103.
60. B. Woods, *Trans. Inst. Metal Finishing, Printed Circuit Supplement*, 1968, p. 134.
61. J. J. Sagura and J. A. van Allen, U.S. Patent 2,940,853 (1960).
62. M. Hepher and H. M. Wagner, British Patent 892,811 (1962).
63. M. Hepher, British Patent 801,735 (1958).
64. G. A. Delzenne, *Tech. Pap., First Reg. Tech. Conf. Soc. Plast. Eng., Mid-Hudson Sec.*, November, 1967, p. 5.
65. D. G. Borden, U.S. Patent 3,169,868 (1965).
66. L. E. Martinson, U.S. Patent 2,697,039 (1954).

67. J. G. Smith and L. M. Minsk, U.S. Patent 2,691,584 (1954).
68. L. E. Martinson, U.S. Patent 3,497,356 (1970) and French Patent 1,528,489 (1968).
69. J. R. Celeste, U.S. Patent 3,469,982 (1969).
70 .F. J. Rauner and L. E. Martinson, U.S. Patent 3,493,380 (1970) and French Patent 1,528,496 (1968).
71. G. Kogel and H. Nevenhaus, U.S. Patent 1,444,469 (1923).
72. R. D. Burr, *Microphotography Symp. Soc. Photographic Sci. Eng.* (Boston chapter), April, 1968.
73. T. L. Steinhoff, C. R. Shipley, and J. A. Macdonald, Canadian Patent 774,047 (1967) and U.S. Patent 3,402,044 (1968).
74. K. H. Saunders, *The Aromatic Diazo Compounds*, Edward Arnold (Publishers) Ltd., London, 1936, p. vii.
75. Kalle Co. AG, British Patent 784,672 (1957).
76. R. G. D. Moore, U.S. Patent 2,797,213 (1957).
77. M. P. Schmidt, U.S. Patent 3,046,111 (1962).
78. W. Neugebauer and F. Endermann, U.S. Patent 3,106,465 (1963).
79. F. Endermann, W. Neugebauer, and M. K. Reichel, U.S. Patent 3,148,983 (1964).
80. M. P. Schmidt, U.S. Patent 3,046,118 (1962).
81. G. Fritz, O. Sus, F. Uhlig, and W. Neugebauer, U.S. Patent 3,184,310 (1965).
82. Kalle Co. AG, British Patent 1,127,996 (1968).
83. K. R. Dunham, D. G. Borden, D. L. Fields, and J. Blair, South African Patent 68 01 224 (1968); also see U.S. Patent 3,567,453 (1970).
84. O. Sus, U.S. Patent 3,046,119 (1962).
85. M. P. Schmidt and O. Sus, U.S. Patent 3,046,120 (1962).
86. Kalle Co. AG, British Patent 711,626 (1954).
87. O. Sus, *Lebig's Annlen der Chemie*, vol. 556, p. 65, 1944.
88. Eastman Kodak, French Patent 1,542,811 (1969).
89. O. Sus and W. Schaefer, U.S. Patent 3,092,494 (1963).
90. J. W. Aylsworth, U.S. Patent 1,029,737 (1912).
91. C. W. Christensen, French Patent 1,542,334 (1968).
92. W. M. Smith, *Vinyl Resins*, Reinhold Publishing Corporation, New York, 1961, pp. 58,59.
93. I. Ostromislensky, U.S. Patent 1,676,281 (1928).
94. T. L. Steinhoff, C. Isaacson, and G. J. Roelants, U.S. Patent 3,535,157 (1970).
95. W. Neugebauer and F. Enderman, U.S. Patent 3,201,239 (1965).
96. H. A. Levine, *Polymer Preprints*, Division of Polymer Chemistry, ACS, vol. 10, no. 1, April, 1969.
97. F. J. Loprest, GAF, South African Patent 70 01 019 (1970).

CHAPTER THREE

Cleaning

After one considers the type of photoresist to use, it is then desirable to perfect an adequate cleaning process and select the related equipment. The importance of cleaning should not be underestimated. Most processing problems can usually be traced to one of two causes—cleaning or lack of process control. Specifically, for photoresist processes cleaning or improper exposure are the usual causes of problems. In this chapter, we will be concerned with cleaning and deal with exposure in chapters on processing individual types of resists.

In spite of its importance, cleaning circuit boards has received rather scant attention. This is due in part to the fact that the soil level on circuit boards is not generally as great as many other forms of metal finishing and due in part to the ability of many photoresists to adhere to somewhat soiled surfaces well enough to get through most plating and etching operations. There is little doubt, however, that many of the problems which come and go without resolution are due to inadequate cleaning. Moreover, subtle losses due to resist failures over a period of time may also be the result of inadequate or improper cleaning.

It is common knowledge that inadequate cleaning results in poor adhesion between the metal surface and the photoresist. Adhesion is generally conceded to be the result of secondary forces that exist between the polymer molecules and polar sites on the metal surface. Polymer molecules are able to orient themselves around these sites for maximum adhesion only when in very close proximity. Films of oil, grease, dirt, or loosely adhering scale prevent the polymer molecules from attaching themselves directly to the metal surface unless the soils are removed by cleaning. Further, defects in the coatings, such as pinholes and poorly wetted uneven resist films, may be caused by particulate matter or surface contamination.

In this chapter, an attempt has been made to assist the reader in the selection of a useful, economical cleaning process prior to resist application. To do this, the nature and type of soils that may be encountered on circuit boards will be pointed out initially followed by a discussion of cleaning sequences and solutions. A final section giving examples of cleaning cycles has also been included.

TYPES OF SOILS

Contaminants can come from a wide variety of sources in a circuit-board process. In some instances, the contamination can reach boards that avoid the contamination process, through various and sometimes devious means. Table 3-1 shows some specific soils with possible sources. The soils have been basically categorized by their nature as organic, particulate, and metallic.

Organic Soils

Organic soils are generally adherent, film-forming, water-repellent, hydrophobic substances. Their presence can prevent the complete removal of other soils, such as oxides, hydroxides, and salts. Perhaps this group is best represented by oils and greases, which seem to have a way of always being present. In fact, nearly every metal-finishing process contains degreasing steps even though it is precautionary in some instances.

Lubricating oils, a common contaminant, usually contain fatty-acid esters of glycerine or hydrocarbon chains as the base substance along with a variety of additives to improve their performance (1). Some additives are listed in Table 3-2. The boundary lubricants and fatty acids can actually react with copper oxides to form a chemical compound tightly bound to the surface. Rust inhibitors also form films on the surface which retard water penetration. Sulfur-containing compounds are quite reactive with copper surfaces also.

TABLE 3-1 Examples of Soils on Circuit Boards

Basic soil type	Specific soils	Sources
Organic	Oils and greases	Drilling or routing equipment, contaminated cleaning solutions, oil from solder reflowing processes, hand lotions, hydraulic fluids from equipment, air compressors
	Fingerprints and sweat marks	Personnel handling
	Resinous and polymeric materials	Mold release agents, laminate bleed-out, oil additives, fluxes, residual photoresists, residual screening and touch-up inks
	Miscellaneous contaminants, wetting agents, foam suppressors	Photoresist strippers, poor rinses from cleaners
Particulate	Copper and plastic board particles, paper, wood, fibers, drilling backup board material, abrasive particles	Drilling or routing operation, personnel clothing, sanding, scrubbing with pumice or abrasives
Metallic	Metal oxides	Baking, heating, air oxidation
	Metal salts	Poor rinsing, wet storage, perspiration, airborne sulfur compounds

TABLE 3-2 Examples of Oil Additives (1)

Additive	Example
1. Antioxidants and metal deactivators	Phenolic compounds
2. Boundary lubricants	Fatty acids (C_{14}) and their soaps
3. Pour-point depressants	Chlorinated wax condensed with naphthalene or phenol
4. Viscosity-index improvers	Polymethyl methacrylate esters
5. Rust inhibitors	Alkyl sulfonic acids and salts
6. Detergents	Alkyl dithiophosphoric acid; salts of calcium, zinc, aluminum, or other elements

Greases are composed primarily of mixtures of soaps and hydrocarbon oils. Examples of soaps are aluminum, barium, calcium, lithium, sodium, lead, and strontium salts of fatty acids. Soaps are dispersed in the oils, which may be paraffins, naphthalums, and other aromatics. Any of the following may be present also (2): (1) fatty esters, (2) fatty acids, (3) caustic compounds, (4) glycerine, (5) fatty alcohols, (6) rosin, (7) water, (8) stabilizers, (9) graphite, and (10) sulfides.

Perspiration, or sweat, contains compounds which are reactive with copper. When placed upon a copper surface by handling, stains that are difficult to remove are formed. Sweat contains sodium and potassium chloride salts, low-molecular-weight aliphatic acids, urea, moisture, and other compounds which strongly promote the corrosion of copper. The corrosion products are hydrophobic and are completely removed only by processes which scrub or etch the base metal, particularly if the part was baked with the fingerprints or sweat in place.

Particulate Matter

Particulate matter is probably the most difficult to define, because there are so many possible sources of particulate matter. However, the particulate matter of most concern in circuit-board cleaning operations is that used in abrasive cleaners, e.g., pumice. These fine particles may be embedded into the surface, adsorbed upon it, or both. Water rinses and blowing with compressed air will only remove the loosely adherent matter and will still leave a thin layer. Embedded or adherent layers of pumice can cause poor resist adhesion, solder dewetting, pinholes, resist breakdown, and poor plating adhesion. Some automated machines follow the scrubbing with a wet brushing to ensure complete removal of the abrasive from the metal surface. Hand scrubbing is not so reliable in general as the machines, and it is necessary to follow the hand cycle with other forms of cleaning to ensure complete removal.

Metal particles from circuit-board machining operations often adhere to the surface but are generally less of a problem than pumice or abrasives.

Metallic Soils

Once free of contamination that can interfere with the functioning of aqueous acid cleaning solutions, metallic soils are relatively easily removed. The oxides, hydroxides, and salts of copper form soluble sulfates, nitrates, and chlorides in acid solutions. Sulfides and fatty-acid salts formed on the copper surface are particularly difficult to remove. Abrasives or etching solutions are frequently employed.

The most common metallic soils are the oxides formed by baking or corrosion products from residual plating salts and poor rinsing and drying.

CLEANING

In this section cleaners and cleaning sequences will be considered. Cleaners will be discussed in relation to the soils which they remove. Table 3-3 shows the cleaning operation required to remove a specific category of soil. Initially, the sequence in which the cleaning operations are applied will be discussed following each of the three individual cleaning operations.

TABLE 3-3 Cleaning Sequence

Step	Soil type removed	Cleaning operation
1.	Organic	Degreasing
2.	Particulate	Particulate-matter removal
3.	Metallic	Acidification

Cleaning Sequence

In the preparation of copper (or any metal) for subsequent processing, it is almost universally accepted that the part will arrive at the cleaning station with some form of grease, oil, or organic contamination on its surface. These substances are sufficiently water-repellent to interfere with the uniform cleaning action of solutions without degreasing capabilities. In other words, they tend to protect other kinds of soils from being removed by virtue of their film-forming, hydrophobic nature. Even though these soils may have no apparent source in your prior processes (other than fingerprints), experience dictates that an allowance be made for their removal in the initial process steps to ensure adequate cleaning by the remaining steps. The formation of a completely wetted copper surface during rinsing usually indicates that the water-repellent soils have been removed.

The process will normally finish with an acid dip to ensure good resist adhesion. Particulate matter may be removed in either the degreasing or acidification processes, or simply by wet brushing if the matter is not present with organic soils. In the following discussion, only organic and metallic soil removal will be considered extensively. Cleaners that remove particulate matter will be noted. Consideration will also be given in the subsequent section on cleaning processes.

Degreasing

Degreasing and the removal of organic soils can be accomplished by one or more of the following cleaners: (1) solvents and solvent emulsions, (2) soak cleaners, (3) electrolytic cleaners, and (4) abrasive cleaners. The actual selection will depend upon the amount and type of soil to be removed.

Solvents and Solvent Emulsions

Solvents and solvent emulsions are unexcelled at removing heavy coatings of mineral and vegetable oils, greases, waxes, and some polymeric substances. They do not remove fingerprints or metallic contaminants, but with proper agitation they can remove particulate matter.

Solvents do not react chemically with the materials they dissolve. Conversely, when the solvents evaporate, a thin layer of the contaminant can be redeposited. To overcome this problem, a final rinse in fresh solvent or in the vapor phase (vapor degreasing) of a boiling solvent is often employed. Herein, a soak cleaner follows all solvent-cleaning steps.

There are many types of solvents, but for degreasing, hydrocarbons or chlorinated hydrocarbons are most often used. (Hydrocarbons are a large group of solvents that contain only carbon and hydrogen atoms.) These solvents are obtained from coal or petroleum and are good solvents for oils. The primary disadvantage is the high flammability and low flash point. Examples include benzene, xylene, toluene, and Stoddard solvent. The last is a mixture of aromatic and aliphatic petroleum hydrocarbons with a distillation range of approximately 300 to 400°F (3), although better grades are available.

Because of the flammability of hydrocarbons, chlorinated hydrocarbons are preferable. These compounds are similar in structure to the hydrocarbons except that one or more of the hydrogen atoms have been replaced by a chlorine atom. The presence of the chlorine atoms improves the solvency range to more polar substances, such as resins, and drastically reduces the flammability. Chlorination also increases the specific gravity and boiling point of the hydrocarbon. Nearly all degreasing is performed in one of three solvents: 1,1,1-trichloroethane, trichloroethylene, and perchloroethylene. The properties of these solvents are shown in Table 3-4. For vapor degreasing of circuit boards, any of the solvents is usable if none of the plastic base material is exposed and immersion times are kept to a minimum. The 1,1,1-trichloroethane usually requires a special water-removal feature on degreasers. Trichloroethylene degreases well but is a known smog producer. Perchloroethylene is an excellent degreasing solvent, but its high boiling temperature can damage heat-sensitive laminates. One excellent degreasing method utilizes a hot ultrasonic soak with perchloroethylene followed by a rinse in the same solvent. A number of solvent grades are available, but degreasing grades are generally satisfactory for most applications (4).

Chlorinated solvents are somewhat toxic, and adequate precautions must be taken. The least toxic in a general sense appears to be 1,1,1-trichloroethane, but it is also the most volatile. Trichloroethylene, as indicated earlier, is a known air pollutant and its use is not recommended. Emulsion cleaners essentially consist of a hydrocarbon solvent containing up to 10

TABLE 3-4 Properties of Degreasing Solvents

Property	Trichloroethylene	Perchloroethylene	1,1,1-Trichloroethane
Molecular weight	131.4	165.8	133.4
Boiling point (°F, °C)	188, 87	250, 121	165, 74
Pounds per gallon	12.2	13.6	10.8
Flash point (open cup)	None	None	None
Fire point (open cup)	None	None	None
Specific gravity			
Liquid	1.46	1.62	1.31
Vapor (air = 1)	4.54	5.83	
Evaporation rate at 77°F (gal/ft² · d)	0.50	0.15	0.60
Allowable concentration in air (TLV) (ppm)	100	100	350

percent emulsified water. Though effective, there is a tendency to leave a film of hydrocarbon solvent and emulsifier, which must be removed by further processing.

Soak Cleaners. The alkaline soak cleaners are probably the oldest method of degreasing. Early solutions were extremely alkaline, and the degreasing action was mainly due to saponification of the fatty acid's esters, which made up the bulk of greases and oils. These cleaners caused the formation of copper oxides and hydroxides because of the high operating temperature and alkalinity, and severely etched light metals. Today, cleaners depend primarily upon emulsification, which is a physical process, rather than saponification, which involves a chemical reaction. Emulsification is a complex subject. At the risk of oversimplification, the emulsification process involves the formation of very fine liquid particles of the soil, which are suspended in the cleaning solution. The usual emulsifying materials present in all modern cleaners are wetting agents and detergents. These compounds contain hydrophobic and hydrophilic groups in the same molecule. The hydrophobic portion of the molecule is attracted to the oils and greases, whereas the hydrophilic end is attracted toward water. To emulsify an oil particle, the hydrophobic ends of the molecule form a shell around the outside surface of the particle with the hydrophilic end outward. This makes the particle sufficiently hydrophilic to exist in the aqueous medium, although the oil molecule remains chemically unchanged. Wetting agents also reduce the surface tension of the cleaner and promote wetting of the soiled metal surfaces.

Saponification of oils, such as the fatty-acid esters of glycerine, which are typical of animal and vegetable oils, requires a pH of 13 or more. The resulting compounds, the alkali-metal salts of the fatty acids and glycerine, are both soluble in water. Emulsification does not require a high pH to remove soils and with the proper wetting agent can in fact take place in acid solutions. Wetting agents are generally classified as "anionic," "cationic," and "nonionic" depending upon the nature of the solution. Anionic wetting agents are used in alkaline solutions. These are usually the alkali-metal salts of sulfonic and carboxylic-acid groups attached to a hydrophobic group, such as a hydrocarbon chain. In acids, the compounds are insolubilized. Cationic wetting agents contain a hydrophobic chain attached to an ammonium group, which forms a soluble salt in acid solutions. Nonionic wetting agents have the advantage of being usable in either acid or alkaline solutions, though they are generally less soluble in alkaline solutions than in acids. Other advantages include low foaming and compatibility with heavy-metal ions, other wetting agents, and hard water. Disadvantages include high cost and their usual liquid form.

In addition to wetting agents, soak cleaners contain a base substance, or "builders" as they are sometimes called. Alkaline-cleaner formulations contain bases that provide a source of available alkalinity and buffering action, soften water, improve rinsability over plain caustics, increase the ability to emulsify oils and greases, and reduce the tendency for soil redeposition. Examples include silicates, phosphates, borates, carbonates, and combinations thereof. Chelating agents for metals and hard-water cations, such as calcium or magnesium, are also considered as builders. Ethylenediaminetetraacetic acid (EDTA) salts are frequently used for this purpose in hand soaps as well as metal cleaners (5).

Perhaps the single most important builder is the silicate group. There are a variety of colloidal silicates with varying degrees of alkalinities. The most commonly used silicate compound is the metasilicate (Na_2SiO_3), or "waterglass." Others include the highly alkaline sesquisilicate ($Na_6Si_2O_7$) and orthosilicate (Na_4SiO_4). Silicates are stable, and they have strong emulsifying power and good wetting properties. In addition, silicates are good buffers, are compatible with other builders, do not attack most metals, and are a source of alkalinity.

Phosphates, particularly complex metaphosphates, are also widely used in alkaline cleaners, although recent studies indicate they are undesirable water pollutants. The colloidal metapolyphosphates are good emulsifiers, rinse cleanly, and have buffering properties. Moreover, these compounds are good water softeners, are compatible with other builders and wetting agents, and are a fairly good source of alkalinity. Sodium pyrophosphate is also a good emulsifier but is expensive. One disadvantage of the pyro- and polyphosphate is their tendency to hydrolyze to the orthophosphate at elevated temperatures and thereby lose their colloidal properties. Orthophosphates, such as trisodium phosphate ($Na_3PO_4 \cdot 12H_2O$), are inexpensive, but they are not good emulsifiers. The orthophosphates, however, can soften water by precipitation of water-hardening cations, such as calcium or magnesium, are a ready source of alkalinity, and can act as a buffering agent.

Carbonates, primarily sodium carbonate, are very inexpensive and a good source of alkalinity. Carbonates rinse more freely than plain caustic. Sodium carbonate has some emulsifing power.

TABLE 3-5 Alkaline Soak Cleaner

Compound	Percent	Function
Sodium hydroxide	20	Builders
Sodium carbonate	18	Builders
Tetrasodium phosphate	20	Builders
Sodium metasilicate	30	Builders
Sodium resinate	5	Inexpensive wetting agent
Anionic detergent	5	Wetting agents and emulsifier
Nonionic surfactant	2	

A typical soak cleaner for copper [given by Spence (6)] is shown in Table 3-5. Soak cleaners should be heated to help liquefy oils and greases for more rapid dissolution. Agitation of the solution is also required for good cleaning. Some soak cleaners contain water-soluble resins, such as sodium carboxy methyl cellulose, to reduce soil redeposition. The resin exists as a colloid with an absorbent, attractive surface to retain soils in solution. Cyanides were used in soak cleaners for copper, but their toxicity and instability have virtually eliminated their use.

Acid soak cleaners are also available. These cleaners utilize wetting agents to emulsify oils and greases and usually contain phosphate builders. There are some advantages to be realized with these cleaners over the alkaline cleaners, such as simultaneous removal of oxides and organic soils. These cleaners are also useful for resist-covered parts, having less effect on resist adhesion than most caustic cleaners. Acid cleaners are, generally speaking, not formulated to remove heavy greases and oils, such as might be expected with alkaline cleaners. Acidity is usually not great enough to remove heavy oxides, and it is usually advisable to follow the cleaner with a strong-acid dip. Rinsing should be thorough to remove wetting agents, which are present in abundance.

Soak cleaners effectively remove greases and oils, but not as well as solvents do. Unlike solvents, they have less tendency to redeposit soils in a hydrophobic state. Cleaners can also be sprayed, but care has to be taken to control foaming. The use of foam suppressors can result in a surface unsuitable for photoresist processing by forming films that are difficult to remove.

Adsorption of the surfactant is a possibility which under certain conditions gives a surface free of water breaks. Fatty-acid soaps can be adsorbed on metal surfaces in the form of a double layer (7). The hydrophobic ends of the molecules are oriented toward each other, while one carboxylic acid group is adsorbed on the metal surface and the other retains water. Silicates appeared useful in obtaining better rinsing under these conditions. Nonionic detergents were not adsorbed on the metal surfaces. Cationic detergents tended to cause poor wetting on negatively charged metal surfaces. The positively charged amine was attracted to the metal surface, which tended to orient its hydrophilic portion outward.

Electrocleaners. As the name implies, in electrocleaning, cleaning is brought about through the application of an electric current. The actual cleaning is due to the scrubbing action of gas generated on the surface of the part, and the applied voltage must be sufficient for electrolysis to occur. Due to the vigorous agitation from the gas evolution, the amount and type of wetting agent are factors in these formulations. High-foaming types, which could retain considerable amounts of hydrogen gas, are usually avoided. A small amount of wetting agent is desirable, however, to reduce evaporative losses by forming a thin surface blanket of foam. The stability of the wetting agents is also a factor. The extreme conditions that exist in these cleaners can decompose some of the highly polar agents and form films which are difficult to remove on the work. Nonionic wetting agents are suitable because of their resistance to the effects of electric current.

Cleaner concentrations vary, but they generally run about 4 to 6 oz/gal for copper. Higher salt concentrations can improve cleaning through greater electric conductivity; however, this increases drag-out losses. Solutions are heated to improve electric conductivity and cleaning. Current is applied during or after immersion, usually at 6 to 8 V dc, with the part as the anode or cathode. Copious gas evolution literally scrubs the surface of the part. Constituents in the cleaner prevent redeposition of the soils, and cleaning is achieved. While cleaning voltages and current densities vary with the immersion time, higher current densities for shorter times are preferable.

Both anodic and cathodic methods can be used effectively for cleaning circuit boards, and each has advantages and drawbacks. It has been popular practice to use anodic and cathodic cleaners in sequence, for cleaning ferrous metals, and copper could be readily cleaned in this manner. Cathodic cleaners evolve twice as much gas (hydrogen) as anodic cleaners and are therefore more efficient:

$$2H_2O \longrightarrow 2H_2(\text{cathode}) + O_2(\text{anode})$$

Copper is practically impervious to hydrogen embrittlement; this eliminates a consideration which is so important with certain types of ferrous alloys. It is also said that nascent alkalis, which are, of course, a powerful cleaning aid for organics, are formed at the metal surface. However, metallic impurities in solution can be reduced and deposited; this can be a major problem.

Anodic cleaners are less efficient than the cathodic type because only half the amount of gas is evolved. In addition, copper is oxidized and an oxide film can be formed. However, anodic cleaning does not tend to redeposit metallic impurities, and nascent oxygen is generated which oxidizes and breaks down many fats and oils. Higher concentrations and current densities can be used to increase oxygen generation and cleaning power.

Cleaners are usually alkaline in nature and usually have a complex phosphate, silicate, or carbonate base with a small amount of a low-foaming wetting agent. A formulation is shown in Table 3-6 from an ASTM specification (8), and there are many excellent proprietary formulations.

Neutral and acid cleaners that utilize higher concentrations of wetting agents and emulsifiers for cleaning combined with conducting salts and acids are also available.

TABLE 3-6 Electrocleaner Formulation

Cleaner	
Sodium carbonate	40 to 50%
Trisodium phosphate	25 to 40%
Sodium hydroxide	10 to 25%
Surfactant	As required for a foam blanket
Electrocleaning solution	
Cleaner	4 to 6 oz/gal
Temperature	140–160 °F
Current density	10–30 A/ft^2
Time	1–3 min

Electrocleaning is a fast, thorough, and inexpensive cleaning method. It is particularly effective at removing particulate matter and organic contaminants, and it is generally an acceptable substitute for abrasive cleaning cycles. However, it is limited to rack processes, unlike abrasive cleaning, which is readily conveyerized.

Abrasive Cleaning

Abrasives are the most common method of removing heavy oxides and inorganic soils. There are essentially three types of cleaning methods: (1) abrasive-water scrubbing, (2) sanding, and (3) air-abrasive blasting. All the methods are adaptable to mechanization or hand operation.

For abrasive-water scrubbing, pumice is most frequently used in a water slurry, although fine silica has been used in acids (9). Often, a wetting agent and a builder are added to the mixture for improved cleaning and removal of organic soils (Table 3-7). Mechanized scrubbers, such as those illustrated in Fig. 3-1, are also available. Although the capital investment is fairly substantial, these machines produce excellent finishes reliably and usually contain built-in wet brushing to ensure removal of particulate water. It is possible for lightly soiled boards to be entirely cleaned in these machines if an acid-abrasive slurry capable of emulsifying light organic soils is employed. A cycle of this type should have thorough internal rinsing and a water-brushing chamber to ensure complete removal of residual abrasive matter.

Pumice is in several ways the ideal abrasive for use with scrubbing cycles. It can be used in either acid or alkaline solutions, is available in the fine grits (3 or 4 F) required for circuit-board cleaning, and is inexpensive. A serious drawback to the use of pumice is the possibility of

TABLE 3-7 Examples of Abrasive Cleaners

Hand abrasive cleaner	
Citric acid	1 to 2 oz/gal
Sodium chloride (salt)	4 to 6 oz/gal
Nonionic wetting agent	0.1 oz/gal

Add liquid to make a slurry with pumice. Scrub with a bristle brush. Scrub with a clean brush and water to remove residual pumice afterward.

Acid machine abrasive cleaners	
Citric acid	7 wt percent
Trisodium phosphate	7 wt percent
Nonionic wetting agent	0.1 wt percent
Pumice—4 F or 3 F	Balance

Use 1.0 to 2.5 lb/gal pH 3 to 4

(a)

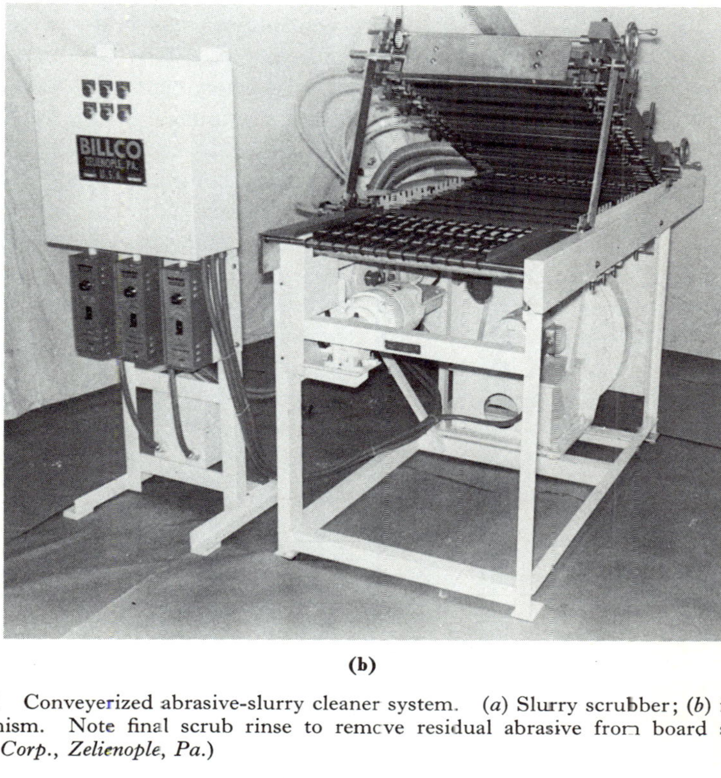

(b)

Fig. 3-1 Conveyerized abrasive-slurry cleaner system. (*a*) Slurry scrubber; (*b*) internal mechanism. Note final scrub rinse to remove residual abrasive from board surface. (*Billco Corp., Zelienople, Pa.*)

embedment into the copper surface. Thwaites and Mackay (10) showed that embedded pumice could interfere with solderability and that this does indeed occur. The amount of embedment, or degree of surface contamination from pumice, was worsened by not using a lubricant (water) or by using heavy pressure during scrubbing. Machines can usually control pressures more uniformly than human operators and are preferable in that respect. Pumice should, however, always be used wet regardless of the mode of operation.

Sanding is another common method of abrasive cleaning. Like abrasive scrubbing, it removes heavy inorganic soils and particulate matter. Abrasives have less tendency to become embedded into the copper surface when attached to a backing, and sanding material can be used wet or dry. Sanding does not remove many organic soils (e.g., oils) as well as wet-abrasive scrub cleaning and should be followed by a soak cleaner as a precaution. Conveyerized sanders do an excellent job of precleaning boards. Care has to be taken to not remove too much copper when sanding; this does not occur with wet-abrasive scrub cleaning.

An intermediate between sanding and scrubbing is offered by machines such as the one illustrated in Fig. 3-2. These machines utilize rotary abrasive wheels of 3M Scotchbrite and similar materials for cleaning. Fresh water is flowed over the part continuously to remove residues. Cleaning of heavy oxides and other metallic residues as well as particulate matter removal is excellent, and the cost of procurement and operation is usually lower than wet-abrasive scrubbing machines. Moreover, metal removal is less than a sanding operation. However, they generally do not remove organic soils, such as oils and greases, completely, and some of these soils can be retained on the wheel. Boards with oils and greases present on the surface should best be solvent cleaned or degreased before cleaning in this manner or sanding.

Airborne abrasive cleaners are not often used as cleaning devices in circuit-board manufacture. With the wide variety of abrasives available, there is no real reason why such a method could not be used, however. The primary reason appears to be the need for specialized equipment. To scrub clean effectively, all that is required are two brushes, a common abrasive, and a tap-water sink. To clean by sanding requires little more. Airborne abrasive cleaning, even in a simple form, requires specialized equipment.

Precautions and Comments on Abrasive Cleaning

Abrasive cleaning mechanically alters the surface of the copper. Methods that can cause deep scratches or that give an overall surface finish of 15 μin or greater should be avoided. The amount of metal removed by abrasive

(a)

(b)

Fig. 3-2 Conveyerized abrasive cleaning unit. (*a*) Abrasive cleaner; (*b*) internal mechanism. (*Billco Corp., Zelienople, Pa.*)

cleaning should be determined before production implementation. If the amount is excessive, adjustments must be made or a different method selected. Care must also be taken to ensure that the cleaning cycle contains provisions to remove residual particulate matter. Rinsing alone is not adequate for abrasive scrubbing. The rinsed surface may appear to be free of particulate matter, but upon drying a visible film is usually left. For adequate removal of the film, the surface should be brushed lightly while still wet with a clean brush or followed by an acid etch, or both. Commercial household cleansers are not recommended for scrubbing circuit boards (11).

Acidification

Acidification is the final step prior to rinsing and drying. This can be accomplished using the solutions shown in Table 3-8 if the oxides and metallic soils are not too heavy and if there is no problem with particulate matter. Often, cleaning cycles containing abrasive cleaning are sufficiently free of heavy oxides to require only a final acid rinse.

TABLE 3-8 Examples of Final Acid Rinses

Acid	Concentration per gallon
Strong-acid rinses	
1. Sulfuric acid	10 vol percent
2. Hydrochloric acid	15 vol percent
3. Perchloric acid	5 vol percent
Organic acids	
4. Oxalic acid	5 oz
5. Sulfamic acid	8 oz
6. Citric acid	6 oz

When heavy oxidation and other metallic soils do require removal, the use of etching solutions prior to the final acid rinse should occur. Examples of some etches are shown in Table 3-9. Perhaps the most frequently used etch is ammonium persulfate at a concentration of 2 lb/gal, where it has an etch rate of about 50 to 100 μin./min. Ammonium persulfate has been shown to leave a film which does not rinse off (12,13) and must be removed by sulfuric acid or other strong acids. Nitric acid etches copper quite rapidly above concentrations of about 35 vol percent, and care must be taken to avoid overetching. Further, copious amounts of toxic nitric oxide are evolved when copper is etched by nitric acid, and

TABLE 3-9 Examples of Acid Etches

Etch composition	Concentration per gallon	Operating temp.
	Ammonium persulfate	
1. Ammonium persulfate	2 lb	Ambient
2. Ammonium persulfate	1 lb	Ambient
Sulfuric acid	10%	
	Nitric acid	
3. Nitric acid	35%	Ambient
Acetic acid	10%	
	Chromates	
4. Chromium trioxide	10 oz	
Sulfuric acid	10%	120°F
Sodium sulfate	2 oz	
5. Chromium trioxide	10 oz	120°F
Nitric acid	10%	
6. Sodium dichromate	8 oz	120°F
Sulfuric acid	10%	
	Ferric oxidizer	
7. Ferric nitrate or sulfate	20 oz	130°F
Sulfuric acid	10%	

adequate ventilation must be provided. Nitric acid solutions tend to form very small pits in the copper surface, which can be eliminated by the addition of acetic acid. In comparison with ammonium persulfate, nitric acid (>35 percent) removes more metal per unit time and is more stable in solutions. Conversely, it tends to pit and evolves toxic fumes, and excessive metal removal is always a serious possibility (0.3 to 1.0 mil/min).

Chromates are powerful oxidizers in concentrated acid solutions and readily etch copper:

$$3Cu + 2HCrO_4^- + 14H^+ \longrightarrow 3Cu^{++} + 2Cr^{+3} + 8H_2O$$

The rate is fairly slow but can readily be increased through the addition of catalytic substances such as halides, except fluorides (14), and alkali-metal sulfates. Etch solutions are stable but must be well ventilated. The etch rate is variable and sensitive to agitation. The use of nitric acid as a source of hydrogen ions tends to give a greater etch rate than sulfuric acid.

Other oxidizers include the ferric ion, cupric chloride, and hydrogen peroxide. Ferric sulfate has long been added to sulfuric acid to aid in the removal of cuprous oxide (Table 3-9). Hydrogen peroxide is unstable and not often used. Cupric and ferric chloride tend to leave a cuprous oxide film on copper surfaces because of the hydrolysis of cuprous chloride formed on the copper surface by both etchants.

There are numerous commercial bright dips for copper utilizing phosphoric, nitric, and acetic acid mixtures. Chromic acid is sometimes added to reduce fuming, and trace amounts of hydrochloric acid may be added for greater polishing action. A basic formulation (15) contains the following:

Phosphoric acid	70 vol percent
Acetic acid	15 vol percent
Nitric acid	15 vol percent

Heating is required for good etching action. These formulations are more expensive than many other etches, but they give a bright, clean surface and have some degreasing capability. They are often sensitive to water content, which can cause pitting. Because of the high phosphoric acid concentration and the nature of the chemical polishing reactions, thorough rinsing is necessary to ensure a clean surface.

Although much of the discussion has centered on the etching cleaners, all must be followed by a strong-acid dip to ensure a clean surface (Table 3-8). Any of the organic acids in Table 3-8 may be used if an etch is not employed and the surface contains only very minor metallic soils, such as water stains. Phosphoric acid and dilute nitric acid solutions have not been included as a final rinse as they may produce undesirable surface finishes.

Rinsing and Drying

The function of rinsing is to reduce the surface contaminants on parts to an insignificant level. To do this, it is necessary that the water itself be fairly pure. Water purity has been the subject of much attention since it was discovered that certain impurities affect the performance of microelectronic devices. Circuit boards do not have the sensitivity to water impurities that microelectronic devices have, but the subject cannot be neglected. In the following section on cleaning cycles, rinses have been classified as "tap" or "deionized" water. Water in the former category refers to that which has been obtained from municipal sources. Tap-water purity varies with the season of the year and with the geographic location. A desirable range of total solids content would be 500 ppm or less with a resistivity of 2,000 Ω·cm or more. The water should be free of turbidity, have a pH range of 6.5 to 7.5, and have a colorless appearance. Deionized water should have a much higher resistivity (50,000 Ω · cm or better) and a much lower total solids content (50 ppm or less). Of course, it would be desirable to utilize deionized water for all rinsing and solution preparation, and there appears to be an industry trend in this direction. Some, however, feel that it is necessary to use tap water where possible for reasons of economy and that with proper rinse placement satisfactory results can be obtained. For this reason, tap water was specified in some instances in the next section on cleaning cycles. For best results, however, the final rinse before drying should always be deionized water.

There are several types of rinses and many considerations in their selection. An extended discussion is well beyond the scope of this book. Good results have been obtained with immersion-spray combinations and thorough spray rinses. In the former system, the part is immersed in an overflowing rinse and given a final spray upon withdrawal. Spray rinsing is in theory the superior rinsing method, but practically there is the danger of incomplete coverage.

Drying is very important in preparing boards for photoresist application. Compressed gases, usually air or sometimes nitrogen, are usually used to force the water from the boards. Commercial conveyerized drying units, such as the one shown in Fig. 3-3, are available. There is the possibility that airborne contaminants, primarily oil and water from the compression machinery, will soil the board, and air line filters are advisable.

Fig. 3-3 Conveyerized drying unit. (*Billco Corp., Zelienople, Pa.*)

CLEANING CYCLES

The goal of the cleaning cycle is clear—it must economically produce a clean, dry, acidified copper surface free of particulate matter to which photoresists can adhere, without removing excessive amounts of metal or damaging the surface. Unfortunately, the words "clean" and "to which photoresists can adhere" represent important factors which are difficult to define accurately. For our purposes, "clean" will simply refer to an acidified metal surface free of water breaks. The latter phrase relates to the use of cleaning methods, particularly in the latter stages of the cycle, which may leave surface films that are not conducive to good photoresist adhesion. Because the latter portion of the cycle always involves acidification, the careful selection of final acid rinses becomes paramount. Resist adhesion is difficult to define quantitatively, and only experience or extensive production testing can be used to truly ascertain the effect on resist adhesion of a given cycle. Fortunately, most modern-day photoresists adhere well to a variety of surfaces, which gives some leeway in the selection of a cycle.

The important factors in the selection of a cleaning cycle are the amount of soil to be removed and the economic considerations, both in capital outlay and in reoccurring material and labor costs.

Five example cleaning cycles have been given to show possible sequences utilizing various innovations and different pieces of equipment. Many other cycles are possible, and a cleaning cycle should depend upon the nature and type of soils to be cleaned. The example cycles are:

1. Hand-abrasive cleaning cycle
2. Short scrubbing machine cycle
3. Sanding-machine cycle
4. Chemical cleaning cycle
5. Heavy-duty cleaning cycle

The first three cycles utilize various abrasive cleaning methods, and the fourth cycle contains no abrasives at all but substitutes electrocleaning. A heavy-duty cleaning cycle has also been included, although the other cycles clean adequately for most purposes.

Hand-Abrasive Cleaning Cycle (Table 3-10)

The hand-abrasive cleaning cycle is perhaps the most common type. It utilizes no special equipment; only a sink, scrub brushes, and tanks are required. Pneumatically operated scrub brushes can be used to speed up the process. As with any scrubbing cycles, metal bristles should not be used. After scrubbing with abrasive, the parts are wet-brushed, or wiped, to remove residual abrasive and racked and soak-cleaned to complete the

TABLE 3-10 Hand-Abrasive Cleaning Cycle

Step	Solution	Process
1. Abrasive scrub	4 F pumice, salt solution (Table 3-7)	Degreasing, heavy particulate matter
2. Rinse		
3. Wet brush	Tap water	Particulate matter
4. Soak clean	Alkaline or acid solution	Degreasing (optional)
5. Rinse	Tap water	
6. Acid dip	Hydrochloric acid	Acidification
7. Final rinse	Deionized water	
8. Dry		

degreasing cycle. A separate clean brush should be used to remove residual abrasive. At this point, the parts should be free of water breaks after rinsing and contain only light oxides. If a neutral- or acid-soak cleaner is used, oxide formation from strongly alkaline cleaners can be avoided. A final acid dip and a thorough rinse complete the process.

Short Scrubbing Machine Cycle (Table 3-11)

The short scrubbing machine cycle is a somewhat unusual cycle in which a conveyerized scrubbing machine, such as the one shown in Fig. 3-1, is utilized. Parts are initially racked and soak-cleaned to remove organic

TABLE 3-11 Short Scrubbing Machine Cycle

Step	Solution	Process
1. Soak clean	Alkaline or acid soak cleaner	Degrease
2. Rinse	Tap water	
3. Acid scrub clean	Abrasive, acid buffer, wetting agent	See Table 3-7 Particulate matter
4. Rinse	Deionized water	Acidification
5. Wet brush	Deionized water	
6. Final rinse	Deionized water	(These steps in the scrub cleaner, see Fig. 3-1)
7. Dry		

contaminants, rinsed thoroughly, and processed through the scrubber. The machine must be charged with an acid type of abrasive cleaner for this process to be usable. Most of these machines are equipped with air knife dryers, which actually reduces the process to two basic steps: soak cleaning and feeding the scrubber. Other forms of degreasing, such as solvent soaking, could be used or perhaps even conveyerized to feed directly into the scrubber for more rapid output.

Sanding-Machine Cycle (Table 3-12)

Although the sanding-machine cycle utilizes a sanding machine as shown, hand wet sanding with pneumatic oscillating sanders could also be used readily. Whether hand or machine sanding, the grit should be 180 grit or finer to ensure a smooth usable finish of 15 μin. or finer. Machine sanding at 180 grit usually gives a smoother surface than can be obtained with hand sanders, and a 280 or 320 grit is preferable for the latter. Hand sanding should be wet to aid in removal of particulate matter. Sanding should be preceded by a degreasing step to remove oils and greases, which would be smeared over the entire surface by the sanding. After the abrasive cleaning, the parts are racked and soak-cleaned to ensure complete degreasing. Rinsing after this step should produce a surface free of water breaks. The only oxides present should be those formed in alkaline soak cleaners. If particulate matter is still present after soak cleaning and rinsing, it is advisable to insert an acid etch before the final acid dip.

TABLE 3-12 Sanding-Machine Cycle

Step	Solution	Process
1. Clean solvent	Chlorinated solvent	Degrease (optional)
2. Abrasive clean	Automatic sander	Heavy particulate matter, metallic soils
3. Soak clean	Cleaner, alkaline or acid	Degrease
4. Rinse	Tap water	
5. Acid dip	Hydrochloric acid	Acidification
6. Final rinse	Deionized water	
7. Dry		

Chemical Cleaning Cycle (Table 3-13)

If abrasive cleaning is undesirable, it is possible to use the chemical scrubbing action of electrocleaning as a substitute in many instances. Even severely soiled copper surfaces can be cleaned adequately without abrasives using a cycle with the combination of electrocleaning and acid etching.

TABLE 3-13 Chemical Cleaning Cycle

Step	Solution	Process
1. Anodic clean	Hot caustic cleaner	Degreasing, particulate matter
2. Rinse	Tap water	
3. Acid etch	Ammonium persulfate	Acidification
4. Rinse	Tap water	
5. Acid dip	Sulfuric acid	
6. Final rinse	Deionized water	
7. Dry		

In the cycle shown in Table 3-13, anodic cleaning is used to scrub off particulate matter and degrease the surface. An oxidized film of copper will be formed in the process. Following rinsing, an etch is used to produce a clean surface free of water breaks. A final acidification completes the cycle. If parts are severely soiled with organics, a solvent cleaning before electrocleaning may be necessary.

Heavy-Duty Cleaning Cycle (Table 3-14)

In the heavy-duty cleaning cycle, a solvent is employed to remove any adsorbed oil or grease on a heavily oxidized surface. The part is then scrubbed to remove the heavy oxide without the danger of smearing or contamination from the oily substances. A soak clean is then employed to remove any residual oil. At this point, the copper surface is free of water breaks, but it may be oxidized lightly. A mild etch of 0.1 mil or less is then employed to clean and remove any residual substances from the surface and to expose an entirely new metal layer; this is followed by an acidification step to complete the process.

This process is more elaborate and expensive than the other processes but is compatible with circuit-board materials. It is intended for parts with heavy baked-on contaminants and severe types of soiling.

Other Metals

Generally, the same concepts with different materials can be followed in cleaning other metals besides copper. A degreasing step, a step in which particulate matter is removed, and the acidification of the surface should all be included. Pumice scrubbing is normally not used as much as when cleaning copper. Table 3-15 gives examples of cleaning methods for some common metals based on the author's experiences. They are to aid in preparing a suitable process. Most of the processes contain etching steps which evolve gases to remove particulate matter. Of course, it

TABLE 3-14 Heavy-Duty Cleaning Cycle

Step	Solution	Process
1. Solvent clean	Chlorinated solvent	Degrease
2. Abrasive scrub	4 F pumice, salt solution (Table 3-7)	Degrease, heavy particulate matter, and metallic soils
3. Rinse	Tap water	
4. Wet brush	Tap water	Particulate matter
5. Rinse	Tap water	
6. Soak clean	Alkaline cleaner	Degrease
7. Rinse	Tap water	
8. Acid etch	Ammonium persulfate–sulfuric acid	
9. Rinse	Tap water	Acidification
10. Acid dip	Sulfuric acid	
11. Final rinse	Deionized water	
12. Dry		

should be realized that the cleaning process depends upon the soils to be removed, and these processes may not be suited for a particular need. It has been assumed that soiling is not extreme and that some etching of the metals is permissible. Extensive soils, such as preserving oils, should be removed by organic solvents before starting the cleaning. Every attempt should be made to have a clean, dry, stable, acidified surface before applying the resist. Conversion coatings are especially useful for many metals, including copper.

Many of the chemicals are toxic and hazardous to handle or otherwise come in contact with and must be used as such. Good ventilation is especially necessary in the etching steps. No immersion times have been given. The times will depend on the extent of the soil. Some typical times are:

Degreasing (except aluminum)	3 to 10 min.
Aluminum etch	20 to 40 s.
Acid etch	10 to 30 s after gas evolution is noted.
Acid dip	Aluminum and stainless steel, 3 to 10 min; others, 1 to 3 min.
Baking	Heat until the part reaches temperature, and hold for 10 to 30 min.
Rinses	60 s or more, depending upon the type.

TABLE 3-15 Some Cleaning Methods for Noncopper Metals and Glass

Metal	Immersion steps (Rinse after each)			Bake	Conversion coating	Remarks
	Degrease	Acid etch	Acid dip			
Aluminum	Alkaline etch cleaner or an etch in 5 oz/gal NaOH at 160°F	....	50% HNO_3	....	Chromate type	Bake after drying if no chromate treatment is used.
Stainless steel	Alkaline soak cleaner	30% HNO_3 5% HF	50% HNO_3 at 120°F	Heat to 120°C	....	Coat immediately after cooling.
Mild steel	Alkaline soak cleaner	50% HCl	5% oxalic	....	Chromate type	
Nickel	Alkaline soak cleaner	40% HNO_3	20% HCl	Heat to 120°C	....	Coat immediately after cooling.
Nichrome alloy	6 oz/gal $KMnO_4$ 6 oz/gal NaOH 140–160°F	3% H_2SO_4 3% H_2O_2	20% HCl	Heat to 120°C	....	Degrease until a thin brown or black smut forms, or 10 min; coat immediately after cooling.
Gold	Spray or ultrasonic solvent cleaner	....	15% H_2SO_4 at 120°F	....	....	Do not rinse after solvent clean.
Glass or quartz	80% H_2SO_4 1 oz/gal CrO_3 80–100°C	....	....	Heat to 200°C (if possible)	....	Coat immediately after cooling.

REFERENCES

1. R. C. Gunderson and A. W. Hart, *Synthetic Lubricants*, Reinhold Publishing Corporation, New York, 1962, pp. 11–25.
2. C. J. Boner, *Manufacture and Application of Lubricating Greases*, Reinhold Publishing Corporation, New York, 1954, p. 11.
3. I. Mellan, *Handbook of Solvents*, vol. I, Reinhold Publishing Corporation, New York, 1957, pp. 113, 118.
4. L. F. Spencer, *Metal Finishing*, June, 1960, p. 55.
5. *Sequestrene Pamphlet*, Alrose Chemical Company, Providence, R.I., 1952, pp. 38–43.
6. L. F. Spencer, *Metal Finishing*, vol. 60, no. 4, p. 63, 1962.
7. K. Durham (ed.), *Surface Activity and Detergency*, Macmillan & Co., Ltd., London, 1961, pp. 78–79.
8. "Recommended Practice for the Preparation of Copper and Copper-based Alloys for Electroplating," ASTM B281-58.
9. H. R. Wollgien and E. H. Chang, U.S. Patent, 3,345,217 (1967).
10. C. J. Thwaites and C. A. Mackay, *Metal Finishing J.*, vol. 14, p. 291, 1968; also see anon., "Tin and Its Uses," no. 81, pp. 3–7, 1969.
11. "Photofabrication of Printed Circuits," *Pamphlet P. 125*, Eastman Kodak Company, p. 11.
12. D. A. Luke, *Trans. Inst. Metal Finishing*, vol. 47, p. 37, 1960.
13. *Circuits Manufacturing*, February, 1970, pp. 42, 43.
14. R. D. Byers and J. G. Poor, U.S. Patent 3,362,911 (1968).
15. *Metal Finishing Guidebook*, Finishing Publishing, Inc., New York, 1957, p. 243; also in years prior and subsequent.

CHAPTER FOUR

Negative Resists

In this chapter, processing details for negative-action wet-film resists will be discussed. Examples of processes will be given for basic resist types along with a discussion of the critical steps. Negative resist resins were discussed in Chap. 2, and their response to light illustrated in Fig. 1-1. As indicated in Chap. 1, negative resists were the earliest photoresists from the standpoint of resists in general and use in the electronics industry. Negative resists are inexpensive to procure; form tough, accurate, adherent images; and are tolerant to variations in processing parameters.

Starting with a basic process, Table 4-1, the individual steps will be discussed in the order shown, along with related information.

RESIST APPLICATION

Coating methods are generally conceded to be the most important consideration in the selection of a process. The fundamental reason is simple. The thickness and uniformity of the coating affect all subsequent processes and the condition of the resist image. Or stated another way, subsequent processing parameters can accommodate certain limited variations in

TABLE 4-1 Basic Negative-Resist Process

Step	Comment
1. Apply resist	Roller-coat, dip, spray, whirl, or flow.
2. Air dry	Expel solvents.
3. Prebake	Expel solvents, harden film, improve adhesion.
4. Expose	Define image, increase adhesion.
5. Develop	Swell and solvate unexposed resist.
6. Dye	Delineate image.
7. Spray rinse	Wash off unexposed resist.
8. Alcohol dip	Sharpen image by shrinkage on some resists.
9. Spray rinse	Wash off remaining residues.
10. Air dry	Remove water, expel solvents, harden image.
11. Examine and touch up	Correct image defects.
12. Postbake	Further harden coating and improve adhesion.

coating thickness. If these limits are exceeded the final-image condition will be affected, usually in an adverse manner. Process parameters, such as bake times and temperatures, may not fully remove the solvents from exceptionally thick coatings, and underexposure will likely occur. Similarly, developing will have more effect on the image, as will dyeing, postbaking, stripping, etc. If the film is too thin, scumming in the form of heat fog, rough images from overexposure, poor dye retention, and a porous, dull final-image finish will result. Process parameters can be adjusted to process either extreme, but not both. It therefore follows that a successful photoresist process must have a coating method that produces reasonably uniform and repeatable resist thicknesses.

How thick and how uniform do resist coatings have to be? An obvious question that is difficult to answer when one considers the variety of uses that are made of resists of this type. As a generalization based upon the writer's experience for the ester resists, a thickness range of 90 to 220μin. is adequate for most etching and plating processes (Table 4-2). Etching processes perform well at the lower end of the range, whereas plating processes generally function best with film thickness of 150 to 220 μin. Cyclized rubber resists can usually be processed slightly thicker than the ester resists, with good results up to 300 μin. or more. Uniformity over the part is also very important in avoiding over- or underprocessing. Quantitative data are scarce with regard to uniformity or to what limits thickness can vary with good ability to be processed. Here again, an accurate absolute answer is very difficult. If measured as the minimum resist thickness over the maximum thickness in the form of percent, uniformity can be expressed as follows:

$$\text{Uniformity} = \frac{\text{minimum thickness}}{\text{maximum thickness}} \times 100$$

$$= \frac{T_{\text{min}}}{T_{\text{max}}} \times 100$$

In this format, values of 60 percent or more can be processed over fairly wide parameters. As indicated earlier, however, it is not an absolute value but one which has proved satisfactory in the author's experience. Thickness measurements may be made using a Microderm or other thickness tester used to measure organic coatings on metals. Measurements should be made in the useful circuit area rather than near the edge unless circuit definition is required there.

TABLE 4-2 Appearance of Films of Resist on Typical Circuit Boards After Processing

Final film thickness (μin.)	Appearance
50	Poor dye retention, spotty appearance with a dull or flat finish
70	Similar except more gloss
100	Glossy dark image that is sensitive to the surface finish of the board
200	Deep reflective gloss uninterrupted by changes in board-surface finish

In addition to thickness and uniformity, other factors must be considered in the selection of a resist application method. These factors and other information are discussed for the following coating methods: (1) roller coating, (2) spraying, (3) dip coating, (4) flow coating, and (5) whirling.

A brief comparison of methods is made in the final portion of this section. Initially, brief discussions of thinning for viscosity control and film formation will be given.

Thinning

To obtain repeatable and uniform coating thicknesses, it is often necessary to adjust and maintain the viscosity of photoresists through the addition of solvents. As a term, "solvent" generally refers to a large group of chemicals that are liquids at room temperature and that are employed to make solids or viscous liquids fluid enough for practical use. Solvents do not react chemically with the materials they dissolve, and they evaporate to allow the formation of a film from the dissolved nonvolatiles. In photoresist systems, there are several types of solvents used, which may be categorized as follows:

1. *True solvents.* As the name implies, these are actually solvents for the resin systems. These are usually the least volatile of the solvent group and are the last to evaporate. One of the more common examples of this type solvent is ethylene glycol monoethyl ether acetate (Cellosolve acetate) for ester resin compositions.
2. *Dilutants.* Liquids in this category are not necessarily solvents for the resin but do not cause precipitation of the resin from the true solvent when added in large quantities. Most photoresist solvent systems are mixtures of the true solvents and dilutants. Aromatic hydrocarbons, such as xylene or toluene, seem to be preferred for dilutants, with chlorinated hydrocarbon being employed when greater solvent action is required.
3. *Thinners.* These liquids are usually the most volatile of the group and are used for viscosity control. When added to the photoresist composition in large quantities, no precipitation should occur. Examples include compatible aliphatic and aromatic hydrocarbons. Dilutants are sometimes employed for thinning resist compositions.

Ideally, each polymer molecule is solvated individually in an envelope of the smaller solvent molecules for the maximum viscosity. Usually, however, small groups or clusters of molecules called "micelles" are solvated along with individual molecules because of the attraction among them, which the solvent is unable to overcome. The use of a poor solvent tends to promote the formation of the polymer clusters rather than solvating the individual molecules and may drastically lower the viscosity or eventually cause precipitation. Therefore, thinning with poor solvents will tend to decrease the viscosity substantially more than thinning with acceptable solvents, such as those recommended by the resist manufacturer. Uneven and defective coatings are the ultimate result of poor thinning practices.

Viscosity can be measured in a number of ways. To be useful as a tool in the control of photoresist thickness, measurement must be resonably accurate and quick to perform, and preferably it must use only a small volume. For more accurate measurements, the Brookfield viscometer gives good results. The comparatively large volume required is inconvenient, however, and this method is perhaps best used for establishing dilution ratios or incoming quality control.

Another method, suggested by Kodak in much of its literature, utilizes the Cannon-Fenske viscometer. This method is accurate and rapid, and requires only a small volume of resist (see ASTM D445-61).

Perhaps the most rapid method for on-site control utilizes the General

Electric Zahn cups. These cups are stainless steel and can be kept in the resist solution when coating by such methods as roller coating or dipping. Operation is simple, requiring only that the time for the cup to empty through an orifice in the base be measured. Usually, less than a minute is required, and several measurements per working day can be conveniently made. More information on this method is contained in the section on roller coating, where the Zahn cups are most often used.

Film Formation and Solvents

Polymer molecules dissolved in a liquid solvent may form a film by virtue of their own size upon evaporation of the solvent. Films may also be formed upon drying by the chemical combination of smaller molecules into larger ones. Most liquid photoresists fall into the former category, which is often referred to as "physical film formation." Film formation by the latter method is quite commonplace and is the method by which many paints and enamels form their adherent films. It is referred to as "chemical film formation" and is obviously undesirable for photoresists which must not be extensively polymerized by air drying because they would become undevelopable.

The term "film" refers to a thin, highly coherent, dry layer of resist. Film formation refers to the transition from the liquid-solution phase to the coherent layer of resist and is a complicated physical process. When drying, the polymer molecules tend to arrange themselves into films, and rapidly evaporating solvents may not allow an even film to form. In photoresist formulations, the true solvent is the least volatile, and when properly diluted for the particular coating method, a wet film with sufficient drying time to allow leveling and surface orientation will be formed. Conversely, solvents that evaporate very slowly are frequently retained by the coatings and may interfere with subsequent processing. (See "Exposure" section in this chapter.)

Roller Coating

In this section, roller coaters suitable for applying uniform coatings of high-viscosity resist will be considered. These were the earliest type of roller coater, and there are some resists that can be used only on these machines. This type of coater has a smooth applicator (or coating) roller and a rotating doctor roller; this type is often used with a conveyerized infrared dryer (Figs. 4-1 and 4-2). As classified in Chap. 5, where three types of coaters are categorized by roller configuration, this is the Type I coater. It will be discussed here because it is generally not satisfactory for applying positive resists. The low viscosity of positive resists yields very thin coatings. In passing, however, it should be noted that the other roller coaters can be used on most, but not all, high-viscosity resists. Generally,

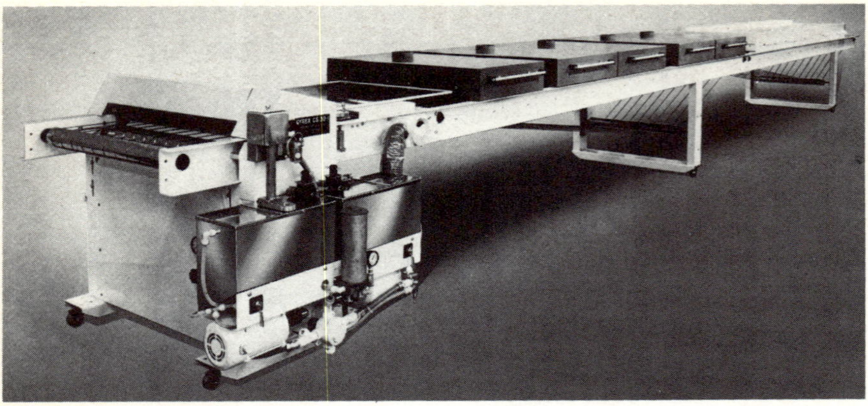

Fig. 4-1 Roller coater with conveyerized infrared dryer. (*Gyrex Corp., Santa Barbara, Calif.*)

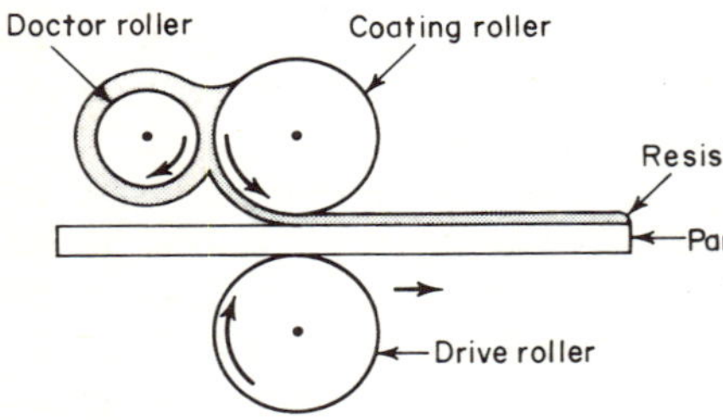

Fig. 4-2 Typical roller-coater mechanism.

high-viscosity resists that produce poor uniformity with the threaded or serrated rollers (Types II and III) can be used with smooth rollers, where less flow is required after coating. Smooth rollers produce much thinner coatings than the Types II and III rollers with the same viscosity resist.

Roller coating, in general, is perhaps the most satisfactory method to apply wet-film resists. Little resist is wasted. It is applied only to the part (and the supporting plate if necessary). In addition to economy, little or no resist reaches the plated through holes by this method, and the output rate is high, although only one side at a time is coated by many machines. Coating uniformity is usually quite good, ranging from 65 to 95 percent and averaging about 80 percent. Roller coaters are easily mechanized and are frequently followed by infrared drying ovens, such as the one shown in Fig. 4-1. Operator dependence is fairly low once an operator has become thoroughly familiar with the equipment, but there is a learning period, particularly if the parts vary considerably in size and thickness. Fairly rough or uneven surfaces can be accommodated but will, of course, produce uneven coatings as with other methods.

Figure 4-2 illustrates the basic operating principle of Type I roller coaters. Resist is flowed between the rotating coating and doctor rolls continuously, working its way to the roll ends; here it is returned to the main sump for filtration and recirculation. Parts are fed between the support and coating roll, where resist is transferred to the upper surface of the panel from the coating roll. Resist is uniformly reapplied to the depleted area of the coating roll by the doctor roll and the entire sequence repeated when another part passes through. Smooth butyl rubber rollers are quite suitable for the ester base resists and solvents, such as Hunt Waycoat RC, Kodak KPR-4, or Dynachem DCR # 3118-H. Dynachem also has a metal-etch-resist roller-coating formulation, CMR 5223. Because of its hydrocarbon solvent system, CMR 5223 cannot be used with butyl rubber rollers.

Several factors affect the thickness of the applied film for a given resist. These factors include (1) conveyer speed, (2) resist viscosity, and (3) doctor-roll pressure.

Conveyer speed has a significant influence on the coating thickness. Increasing the speed of the coating roll increases the thickness deposited on the part. Roughly, doubling the speed results in a 20 to 30 percent film thickness increase. Most roller coaters range in speed from 4 to 18 ft/min and operate well in the middle and upper ranges (8 to 14 ft/min). Resist manufacturers may have specific suggestions for a specific roller coater. Kodak (1) indicated 6.5 ft/min for KPR-4 on the Gyrex 630 roller coater, and Hunt (2) 8 ft/min on the same machine for their Waycoat RC resist. The author's experience favors higher speeds (12 ft/min) on the Gyrex 630 for these resists. Dynachem makes no specific speed recommendations but suggests the instructions of the roller-coater manufacturer be followed.

Viscosity is the primary resist control parameter and variable in this coating system. The resist film is continually exposed to air on the rollers, and solvents evaporate. When this happens, the resist viscosity increases; and a heavy, tacky coating is deposited, with streaks on the parts frequently occurring. Thinning must be done with the manufacturer's recommended solvent system to avoid problems, such as precipitation, unusually large variations in viscosity, or damage to the rollers. Viscosity is controlled from the sump of the coater, where the solvent additions are made. A most convenient method utilizes small stainless-steel cups with orifices of varying size. The cups have long handles and can be conveniently immersed in the resist (or even left in the sump) and the time measured for most of the resist to flow from the cup. Viscosity is determined relative to the time and orifice size. One of the most widely used viscometers for this is the Zahn cup, which is used in the paint industry. Cup orifices sufficient to measure medium oils and mixed paints in the range of 170 to

550 cP in 25 to 60 s (#3 cup, 0.148-in. orifice) are best suited for photoresist applications, although the cups are available in a variety of sizes, including a size for measuring solvent viscosity. It is important that the temperature be held within certain limits during the measurements. For this reason, a stainless-steel dial thermometer permanently inserted in the sump can reduce errors from inaccurate dilutions once the equilibrium temperature of the resist has been established. Resists such as KPR-4, Waycoat RC, or CMR 5223 can operate well around a viscosity of 400 cP, and the Dynachem #3118-H resist near its maximum viscosity. It should be pointed out that while the Zahn cup measurement is related to the viscosity of a liquid, it cannot be directly coverted except for a given type of resist. For example, KPR-4 can be measured using a Zahn cup, and the time related to the absolute viscosity as measured by another method. The temperature and the kind of resist must be the same for good reciprocity.

When one measures the viscosity, care should be taken to accurately time the draining of resist from the cup. A stopwatch is highly desirable but not essential. Timing should start immediately when the cup breaks the fluid surface and end with the first sharp break in the stream of resist draining from the cup. The break is usually quite abrupt and occurs as the stream begins to narrow from the head loss in the cup. It will frequently be followed by dripping, which should be disregarded. Two measurements are usually adequate if they are within 2 s of each other. Control limits of ± 5 s allow a variation in viscosity of approximately 350 to 440 cP for the higher-viscosity ester types of resists, which generally give acceptable coating thicknesses for many uses.

It is good practice to double wet-coat these resists. The operation is quite simple. Parts that have passed through the rollers are immediately returned and a second coat applied after the part is rotated 90°. Double coating, though an additional step, has two distinct advantages: (1) The second coat covers defects that have been missed on the first coat; (2) the second coat increases the thickness by 30 to 40 percent or more, depending upon the type of roller, conveyer speed, resist viscosity, and doctor-roll pressure. Both factors result in less defects and higher yields. However, for volume production double coating may not be feasible.

The doctor-roll pressure has some influence on the coating thickness, particularly with the serrated roller. Coating thickness increases with decreasing pressure regardless of the type of roller. For the smooth roller, a pressure of 15 to 20 psig is adequate for the Gyrex 630, but it should be balanced with the other parameters.

Production models of roller coaters that simultaneously coat both sides of the part during a single pass are now available, e.g., the Gyrex 20. Operational details are not as well defined as for the single-sided coaters, however.

When coating thin or flimsy parts, it is necessary to use a backup plate. Several comments should be made:

1. The plate should be resistant to cleaning solvents and the resist. It should not deteriorate in such a manner as to contaminate or impregnate the dry- or wet-resist film.
2. Nonabsorbent material should be used to avoid retention of resist and cleaning solvents.
3. Surfaces should be flat and very smooth to accommodate resist-coated parts without damage to the resist film.
4. The plate surface should be rigid and heat-resistant.
5. Dimensions should be approximately the same as the parts being processed to avoid part deformation or the application of resist to the plate. The thickness should represent that of the bulk of the rigid parts normally processed, for example, 0.062 in. thick. Thin parts can then be processed, along with the standard parts, without changing the interference settings.

Several materials meet the aforementioned criteria. One of the more suitable is phenolic-glass laminates. Regardless of the material, the surface of the plates should be regularly cleaned and inspected. When using serrated rollers, it is desirable if not essential to use a soft backing on the backup plate in order to reduce the tendency to deposit resist into open-plated through holes or other recesses.

Interference settings of 20 to 30 mils give adequate results. Thinner parts may require smaller interference settings to prevent possible wetting of the driver roller. Large interference settings will tend to wear the coating roll unnecessarily. Also, the machine manufacturer will have recommendations.

Spraying

Spraying has several advantages over other methods. Rough parts, uneven parts, or very large parts can be readily coated by hand spraying. Boards with plated through holes can be coated without clogging or excessively coating the hole walls. On the other hand, operator dependence is somewhat higher than with other methods, and the amount of resist lost because of overspray is significant.

Equipment requirements for hand spray operations are minimal compared with other methods with similar production outputs. Spray booths and guns are reliable pieces of equipment and are available from several suppliers at reasonable cost. Clean, filtered compressed air, or preferably nitrogen, to function as the propellant must be available at regulated pressure. The parts should be clean, dry, and free of particulate matter. Mountings should be provided in the hood to maintain the part horizontal or within 30° thereof. Provisions must also be made to move the parts to another area for air drying in a horizontal position. Metal racks that contact parts only in the extremities of the part are useful and easily cleaned. Both the conventional and so-called "airless" spray guns can be used, depending upon the operator preference, but the gun should be clean prior to filling with resist. Dried resist particles can become loosened and deposit on the part. Such particles can cause spots in the unexposed areas after developing, can clog nozzles, and can cause distorted spray patterns. The particulate matter is in a position where filters cannot remove it and may give the appearance of contaminated unfiltered resist.

Actual application techniques will vary, as indicated earlier. One useful method requires coating the part with a series of evenly spaced vertical passes followed by a similar number of horizontal passes to ensure complete coverage. Depending upon the gun, a pressure of about 20 psig over a distance of 10 to 12 in. should be used. Coverage of a swath about 3 in. wide per pass should be allowed. On an 8- × 10-in. part, for example, four vertical and three horizontal passes are adequate to give good coverage. If the part is not sprayed while flat, it should be allowed to air-dry in this position to ensure good coating uniformity.

The desired condition of the resist immediately after spraying is a smooth wet film with some leveling characteristics. Sprayed resist is actually applied as tiny droplets which flow upon impact with the surface. If the droplets do not contain enough solvents, they cannot flow and level properly, and uneven films result. When too much solvent is used, a thin coating which has a high incidence of defects and may be excessively thick around the edges is frequently obtained.

Spraying covers only one side at a time. Between the front and back coating, the part should be dried well enough to allow some handling without damage or distortion of the film. When spraying the second or back side, care must be taken not to wet the front side with overspray.

Because of operator preference, the amount of thinning required to obtain a suitable coating thickness will vary. For the same reason, dilution ratios of resists and thinners will also vary. A good starting point for thinning is to obtain a viscosity of 10 to 20 cP for most resists. Care should be taken to use the proper solvent for dilution when spraying. Most manufacturers of resists have recommended thinning solvents which should be used unless equivalent solvents are known. The use of improper solvents for thinning can cause poor surface wetting, thin spots in the coating, "orange peel," and other defects. (See "Thinning" section in this chapter.)

Dip Coating

Within certain limits, dip coating is an excellent method of applying photoresists of all types. Both sides are simultaneously coated during the operation, unlike most other methods. However, open holes tend to streak, and nonuniform coatings can frequently occur. Resist losses are also fairly high, depending upon the time the part is allowed to drain back into the sump.

The operation per se is simple and can be performed with no equipment other than an open container of resist. Coating thickness is a function of the withdrawal rate of the part from the resist sump and the viscosity of the resist. Increasing withdrawal rates, usually around 10 or 15 in/min, gives thicker coatings up to a point. Thereafter, the average coating thickness changes very little. Figure 4-3 shows the typical variation curve

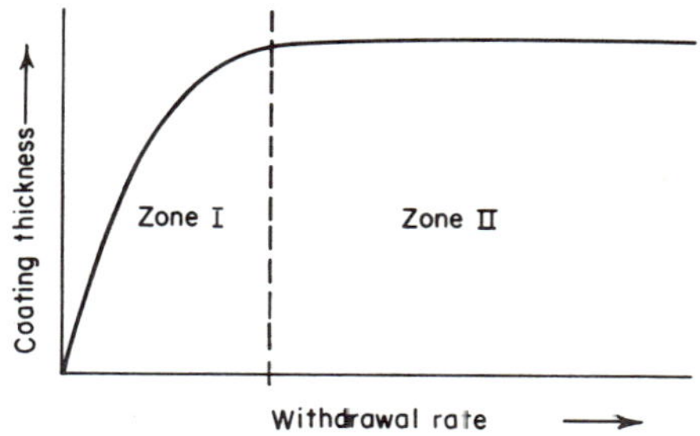

Fig. 4-3 Dip coating. Effect of withdrawal rate on the average coating thickness.

of resist thickness and withdrawal rate. Zone I of the figure, the area wherein the withdrawal rate effects the coating thickness, is where much of the automatic equipment for controlling withdrawal rates operates. Generally, coating uniformity is better in Zone I, but satisfactory coating uniformity can be obtained in either zone (50 to 70 percent is common).

Higher resist viscosities give thicker coatings, as indicated in Fig. 5-3. However, resists with solvents that dry very slowly can produce coatings with poor uniformity, for example, 30 percent, because of a continued, prolonged draining and flow. Resists such as DCR#3116, KPR-3, or Waycoat 20 have good properties for dipping in the as-received condition. Very thick coatings for deep etching can be obtained using dip coating with very viscous solutions of metal-etch resist (3) (150–450 cSt viscosity). Withdrawal rates of less than approximately 3 in/min are essential to maintain the coating factors greater than 50 percent.

Because the resist comes in contact with many different parts, contaminants may tend to build in the resist with use. Particles of board material and other solids must be removed by filtration. Another contaminant is dissolved metals. Most of the resists are acids and capable of dissolving oxides of metals, which can eventually darken or discolor the resist and interfere with exposure. When this happens, the resist must be discarded.

Flow Coating

Flow coating when done by hand is the simplest method of application. It simply consists of pouring a pool of resist on the part and tilting until all areas are wetted and covered. The part is then drained of excess resist for a few seconds by hanging vertically; then it is allowed to dry, preferably horizontally. Obvious disadvantages include problems with uneven coatings and plated through holes, a comparatively high resist usage rate, and high operator dependence. However, equipment costs are incomparably low for hand flowing. Only an area where the parts can be drained and allowed to dry tackfree without becoming contaminated with airborne particulate matter is required. Because the resist is used only once, it is clean and fresh when flowed on the part.

Resists used for flow coating should not be viscous or fast drying. A viscosity range similar to that used for spraying gives acceptable results (10 to 20 cP) or at least a starting point. Drying rates are also acceptable when thinned with the thinners available from the resist manufacturer for spraying.

Commercial flow coaters are available for coating sheet material. The resist is pumped through orifices which flow directly on both sides of the part as it passes through the unit in a vertical position. Excess resist is allowed to drain back into a sump for filtration and recirculation.

Whirling

Although generally reserved for microelectronic applications, whirling has long been used in the graphic arts as a means of coating. The primary reasons against using whirling for circuit boards are the inability to handle parts with through holes and the low output per system compared with other application methods. Parts with holes tend to streak resist in radials outward on the bottom side. Output per whirler is much less than for a spray system or roller coater. Still, whirlers such as those used in the graphic arts can apply coatings that are outstanding in uniformity (90 percent is common) to flat surfaces with no unfilled holes. Any resist can be coated by whirling on parts of nearly any size. Whirlers can also be used for developing. The developing solution is flowed on the part while it is whirling at low speed. However, like spraying, the resist (or developer) can be used only once and considerable losses occur. Less coating uniformity is generally obtained with rectangular parts than with square or round parts, but rarely is less than 80 percent encountered.

Speeds on the macrowhirlers are much slower than those used on the microelectronic variety. Whereas the latter spins at 8,000 rpm or more, 200 rpm is not an uncommon top speed for the former. For resists of the type under discussion, a viscosity of 20 cP at a speed of 100 rpm will give good results with single or multiple coatings. Speeds as low as 75 or as high as 150 rpm have been used with good results. Higher viscosities can be used at the higher revolutions per minute, and vice versa. Two methods of applying the resist to the part can be used. A method commonly used with dichromated coatings is to pour on the resist while the part is whirling. Pouring starts at the center and proceeds outward; excessive resist is thrown from the plate and good distribution is obtained. Frequently, two coats are applied to ensure complete coverage. The second method consists of flooding the part immediately before starting the whirling; wet spraying is sometimes used. A swab or brush is also useful to ensure wetting of all areas with the latter method. Whirling must start immediately, while the layer is still fully liquid. As may be surmised, a fairly large quantity of resist is used in this method of coating, and most of it is whirled off the part. Although this is no great problem for the users of dichromated resists which are much less expensive than the resists under discussion, it certainly adds up when used with volume production of large parts.

AIR DRYING

Air drying is an important step for some types of resists, less so for others. It appears helpful for high-molecular-weight resin systems that contain a combination of highly volatile solvents which function as a dilutant and a true solvent that is much less volatile. After resist application, rapid forced-air drying can form a surface skim with the evaporation of the more volatile solvent components. Some of the less volatile true solvent can be retained beneath the surface skim. The resulting coating may appear hard and adequately processed at this point. Upon exposure, however, the retained solvents interfere with polymerization. During developing, solvation of the partially polymerized resist would be increased, and sloughing off or blistering of the image could occur. The symptoms would be those of underexposure, poor cleaning, or both. Because the times after coating and before prebaking may vary, the problem could come and go somewhat mysteriously. Thick coatings have a much greater tendency to retain solvents than thinner ones, and lower-molecular-weight resins, such as Dynachem DCR or Kodak KPR, have less tendency toward skimming than higher-molecular-weight systems, such as those in KPR-2 or Hunt Waycoat 20.

The coating method and type of prebake also can affect the tendency to skim. Dip coatings are drained and air-dried during the process, and roller-coated resist layers contain little volatile solvent. Spray coatings usually contain large amounts of volatiles but are allowed a generous air drying before prebaking because of the tendency of the resist to run and form droplets. Whirling usually continues until the coating is dry, and some whirlers have built-in heaters for prebaking. Flow-coated parts are allowed to drain before prebaking. As may be seen, then, most coating processes have built-in air-drying operations. The problems stem from the fact that the duration of the air-dry step is sometimes too short for some resists.

Infrared prebaking virtually eliminates the possibility of skimming by rapidly heating the copper surface preferentially and expelling the solvents from within, whereas air convection ovens tend to heat the resist surface initially, which is a primary cause of skimming. Air drying is usually unnecessary with an infrared prebake step.

The air-drying step does not have to be extensive in duration. Drying tackfree is preferable but a slight tackiness is acceptable. Under average conditions, 5 to 10 min of air drying is adequate.

PREBAKING

Preexposure baking, or simply prebaking as it is preferably termed, follows the air-drying step in the basic process. It has three basic purposes:

1. To expel retained solvents to ensure repeatable exposure times
2. To harden the film to resist damage from routine handling and developing
3. To improve the adhesion of the coating to the copper surface to better withstand the developing step

Prebaking is an essential step in the process. Most resists have a fairly wide latitude on prebaking temperature and times. If the temperature and/or time are excessive, thermal polymerization can occur, which results in loss of photosensitivity or scum formation ("heat fog"). When the cycle is inadequate, erratic exposure, poor adhesion, and image damage in the developer can occur. It is desirable in the processing prior to exposure to have the coating in such a condition that exposure will create the maximum difference between the image and nonimage areas. To do this, the prebake would be adequate to expel the solvents and harden the film, but not enough to cause any significant degree of cross linking. This condition would ensure proper and repeatable exposure of the image areas, which would provide adequate resistance to the developing solution. Unexposed areas should be readily swelled and solvated for ease of removal by the developing step.

Most resists can be prebaked over a fairly wide range of times and temperatures. Shorter times at higher temperatures or longer times at lower temperatures can be used interchangeably. Figure 4-4 shows a prebake envelope that has been used for KPR-2, a very versatile resist. Some resists, such as KOR, are much more sensitive than KPR-2 to thermal polymerization and would have a much different envelope. Table 4-3 gives prebake parameters for some resists.

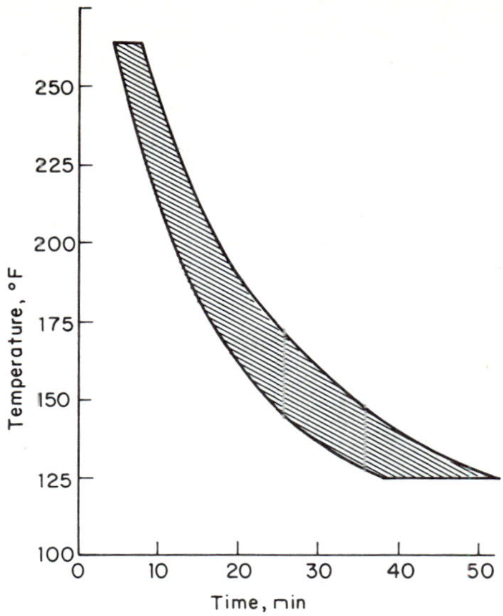

Fig. 4-4 Some pre bake times and temperatures that have been used for KPR-2.

TABLE 4-3 Typical Preexposure Baking Parameters for Some Resists

Resist	Temperature (°F)	Time (min)
KPR, KPR-2, KPR-3	180	15–20
KPR-4	220	20–25
KOR	150	5–10
Waycoat RC	200	20–25
Waycoat 20	180	15–20
DCR #3118	140	10–15
DCR #3140	125	10–15
KMER	180	20–25
Waycoat PF	150	20–25
CMR-5223	150	20–25

EXPOSURE

Adequate exposure and the factors affecting it are the most serious problem source in the processing of negative-acting resists. Exposure creates the difference in the coated areas that form the final image. As explained in Chap. 2, exposure brings about further polymerization of the polymer molecules, which imparts greater chemical resistance and adhesion. Inadequate exposure tends to manifest itself in soft, excessively swollen, poorly adherent images that are damaged or lost in the developing or subsequent spray rinsing operation. Sometimes, if the exposure is marginal, the image will endure the processing but have a dull, nonreflective finish instead of the glossy, lacquerlike finish that should be obtained.

Photoresist sensitometry is a complex subject when considered from a theoretical viewpoint, but it does lend itself to some practical approaches. Films of resists are made up of a quantity of photosensitive materials, and a related quantitative amount of radiation must be absorbed and reacted for insolubilization to occur. Only the portion of the radiation that is absorbed can react, and not all the reactions that occur are fruitful toward insolubilization. Absorption occurs only in a narrow frequency band that corresponds to a wavelength of 2000 to 5000 Å for most resists. Figure 4-5 illustrates the portion of the radiation spectrum of interest when exposing photoresists. As may be seen, some of the visible spectrum which

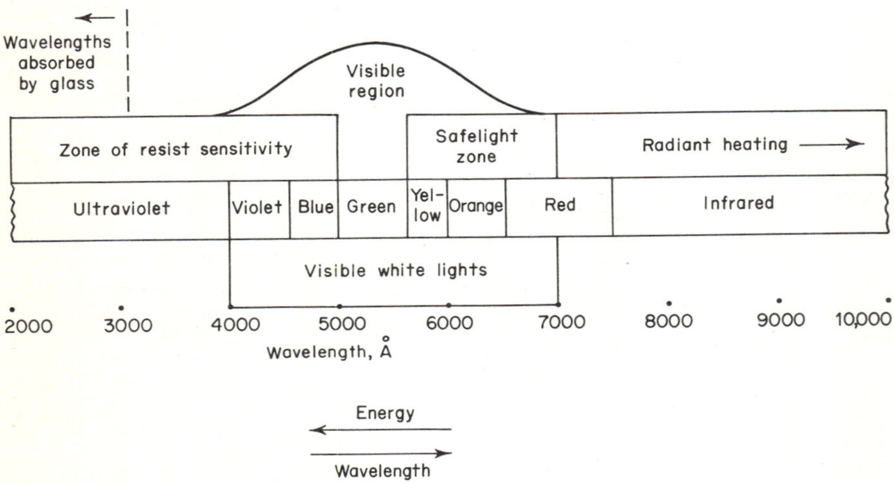

Fig. 4-5 Photoresist spectrum.

is also present in white light overlaps the photoresist sensitivity zone. To be safe, it is desirable to work in the yellow zone unless the resist sensitivity extends further into the visible range, for example, KOR, in which case orange or red light is used. Needless to say, orange- and red-light sources can be used with all common photoresists. Ultraviolet-light sources produce visible and invisible radiation as a general rule, and the latter is lost first when the source begins to deteriorate. Because the invisible high-energy ultraviolet radiation produces the resist insolubilization, it is quite possible for a light source to "look" all right but give poor exposures. It is not intended that one should look directly into an ultraviolet-light source but rather that one refer to the overall appearance reflected from the surroundings.

Perhaps the most widely used light source is the high-pressure mercury-vapor lamp. Radiation from this source is rich in the ultraviolet wavelengths as well as the visible range. Figure 4-6 shows how some of the emission spectrum of mercury vapor extends over the regions of resist sensitivity. Mercury-vapor lamps and related electrical equipment have a long useful life and are readily procurable at reasonable cost. Many types of commercial printers are available that utilize these lamps. Generally,

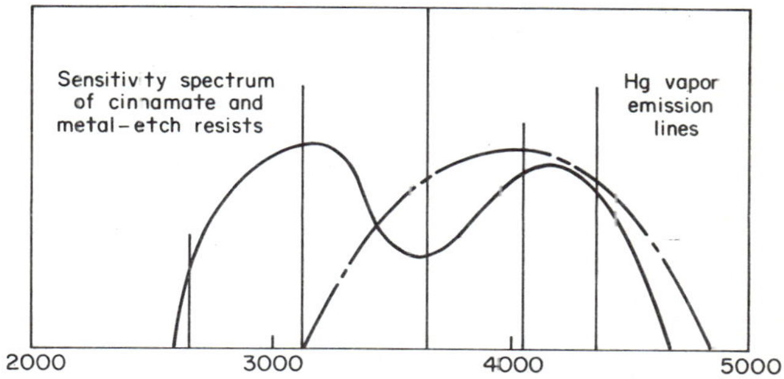

Fig. 4-6 Sensitivity of some negative photoresists shown in relationship to mercury-vapor emission lines.

the printers may be categorized as either fixed or moving light-source types. In the former category are printers such as the ones shown in Figs. 4-7 and 4-8, in which a vacuum frame is inserted between two fixed light sources during printing. The frame is automatically ejected upon expiration of the exposure time preset on the timer. The distance between the lamps and the work varies with the power of the lamp. Considerable infrared radiation is evolved by the high-pressure mercury-vapor lamps, and heat polymerization can occur in unexposed areas if the lamps are too close to the frame.

A fixed 1,000-W lamp in a reflector within the confines of the printer is usually safe at about 30 or 40 in. for most resist processes. More powerful lamps in closer proximity can be used safely if the exposure time is accurately controlled. This is accomplished with moving-light-source or scanning printers such as those shown in Figs. 4-9 and 4-10. In these systems, large (3,000- to 5,000-W) lamps come within 6 in. or less of the frame during exposure. Long tubular lamps are employed, which produces a broad-light-source effect. With some form of collimation, considerable pattern distortion can occur, although the problem is less serious with wet-film resists than with the dry films (Chap. 6). Good scanning printers are equipped with collimators in some form. Collimators are usually not used for the fixed-light-source printers, where the small lamps tend to give a point-source effect. Mercury-vapor lamps require a

Fig. 4-7 Double-sided fixed-light-source printer, the Colight DMVL. (*Colight, Inc., Minneapolis, Minn.*)

warmup after starting until full intensity is reached. Times range from 5 to 30 min.

Other sources employed in printers include the pulsed xenon lamps, which produce radiation very rich in the ultraviolet wavelengths. These lamps are capable of instant starting and stopping, unlike mercury-vapor lamps. Considerable heat is radiated, and the lamps are usually off when not in use because there is no warm-up required. The lamps are small enough to approximate a point source with common power consumption levels of 2,000 to 8,000 W.

Carbon-arc lamps have long been used to print screen stencils and chromated colloid coatings and are quite effective as a light source for negative resists. An arc is produced between carbon electrodes, which are consumed in the process. Carbon-arc printers are equipped with a motor-drive mechanism which maintains the arc length by feeding the carbon electrodes as they are consumed. Disadvantages include the evolution of gases and carbon particles from the electrodes and possible interference with electrical equipment. Carbon electrodes often contain

Fig. 4-8 Table-model fixed-light-source printer, the Colight M-218. (*Colight, Inc., Minneapolis, Minn.*)

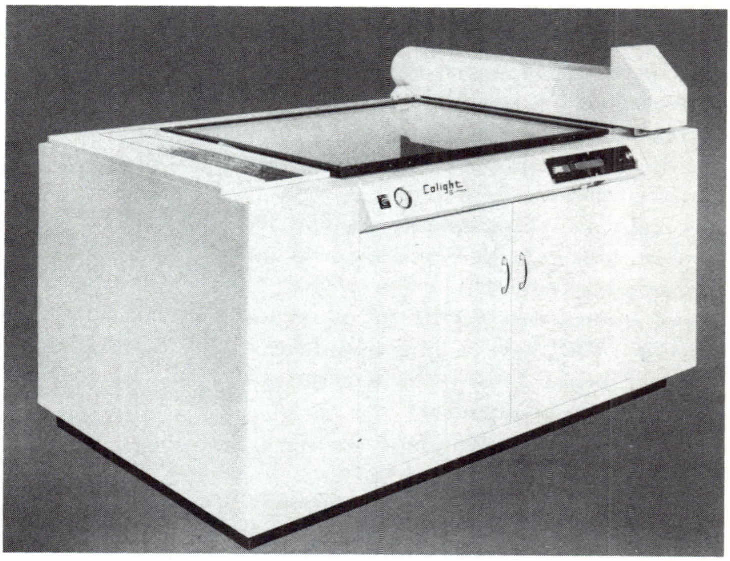

Fig. 4-9 Scanning-light-source printer, the Scanex II. (*Colight, Inc., Minneapolis, Minn.*)

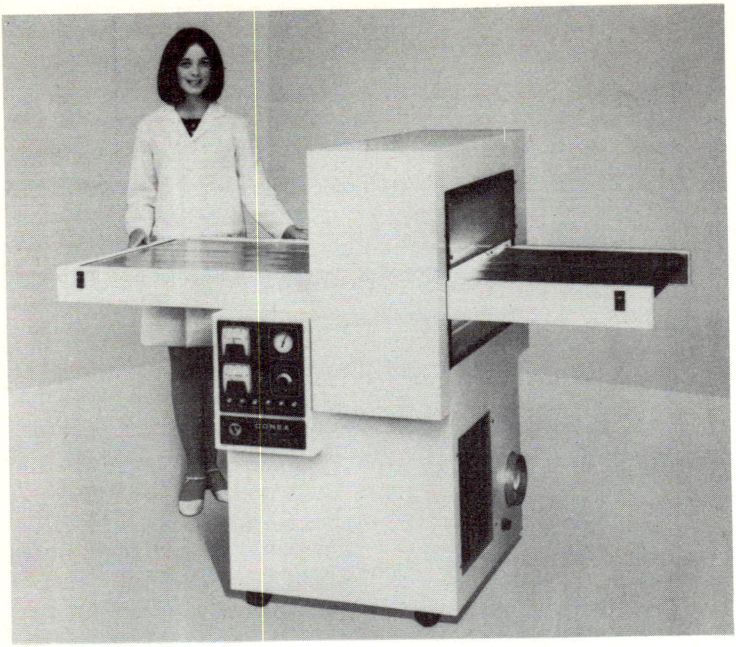

Fig. 4-10 Conveyerized photoresist printer, the Conex System. (*Gyrex Corp., Santa Barbara, Calif.*)

metallic salts to alter the color of the light, which is rich in the violet range of the visible spectrum. Copper salts in the electrodes give a bluer color, which is desirable for some lighting applications, and rare-earth salts, such as cerium fluoride, are used for whiter lights. Carbon electrodes are very effective and inexpensive and can produce excellent images with any commercial photoresist.

Another light source commonly used in fixed-light-source printers is the fluorescent tube. Little heat is evolved, and the distance between the source and printing frame can be 6 in. or less. These lamps may be turned on and used immediately with little or no warm-up, or burned continuously, because they have a long useful life. The disadvantage of the lamps is their broad source characteristics, but for macroapplications excellent results can be obtained.

Before we close this portion of the discussion, a mention should be made of vacuum frames. Poor contact between the photomask and resist film results in a distorted image. This is normally prevented by the use of a vacuum frame. Vacuum frames have a rigid and a flexible side, either or both of which may be transparent. During operation, a temporary seal which forces the rigid and flexible sides against each other and any parts between is made by means of a vacuum pump. Vacuums may vary from about 20 to 28 inHg (1 atm = 29.2 inHg). For most circuit-board work, vacuum frames with a rigid glass or plastic base and a flexible plastic overlay are used (Fig. 4-7). Single sided frames generally employ glass overlays with a flexible grooved or serrated rubber base. Drawing a vacuum forces the part and photomask against the glass for exposure. When exceptionally good contact between the photomask and resist is necessary, bleeder lines in the vacuum frames can be used. Bleeder lines, or "bleeders" as they are sometimes called, start at the vacuum port within the frame and go around the parts in close proximity. The purpose of this innovation is to avoid air entrapment around the part which prevents good contact with the photomask. Bleeders can be made from noncontaminating materials that have no sharp edges to damage the plastic overlay, such as fine plastic-coated wire or nylon cord taped to the base. Grooves in the nonimage area of the rigid base can also be used. For the negative or positive wet-film resists, bleeders are usually unnecessary except for fine-line work. However, they are very useful with the dry-film resists.

During the previous and forthcoming discussions, a very basic and obvious assumption has been made. The image or emulsion side of the photomask should always be in contact with the resist coating during exposure. Otherwise, an adequately exposed but distorted image can result.

Method of Determining Exposure

Exposure time for a given light source controls the quantity of light energy striking the photoresist. It should be determined for each light source–photoresist combination using one of several methods. Although the accuracies of the methods will vary, most photoresists in this category have a fairly wide tolerance for overexposure. Methods of determining exposure times are useful for evaluating equipment, for evaluating photoresists, and for day-to-day process controls. The following methods will be discussed: (1) step-tablet, (2) calculation, (3) inspection, and (4) image-accuracy. The last method includes actual line-width measurements, resolution patterns, and a modified use of the step tablet.

It is very important that all other processing parameters be held constant when varying the exposure time.

Step-Tablet Method. Nearly every manufacturer of photoresists suggests the use of step tablets for macroapplications. The two step tablets most often used are the 21-step Kodak No. 2 and Stouffer 21-step Sensitivity Guide (Fig. 4-11). A "step tablet" has a series of zones or steps in which the optical density increases uniformly by a set increment from step to step. For a given exposure time, the quantity of light transmitted by each step decreases as the density increases. In other words, the

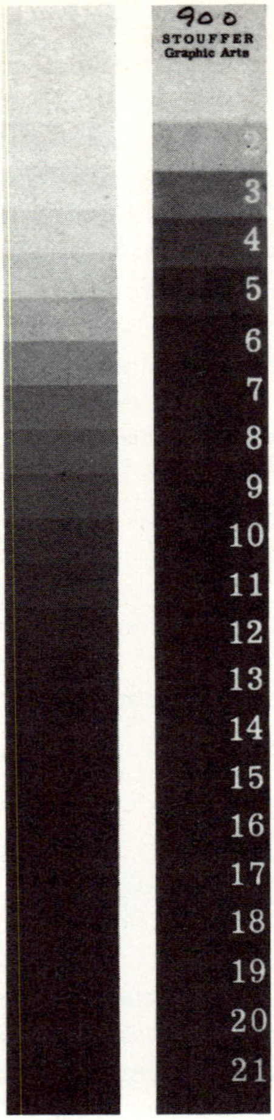

(a) (b)

Fig. 4-11 Step tablets commonly used in photoresist exposure. (*a*) Kodak No. 2 step tablet (*Eastman Kodak Co.*); (*b*) Stouffer 21-Step Sensitivity Guide. (*Stouffer Graphic Arts Equipment Co., South Bend, Ind.*)

photoresist under each step will receive a different quantity of light energy (Fig. 4-12). Increasing the exposure time will increase the number of steps that endure the subsequent developing step. While the condition of the last, or "threshold," step does not reflect the condition of the resist that is not under the step tablet, it is a visual measure of the quantity of light received. Exposure times that will insolubilize the first six steps are adequate for most resists in this category, with allowable times up to eight. The condition of the threshold step may vary and make an accurate determination difficult. As a general rule, if more than half of the area in the

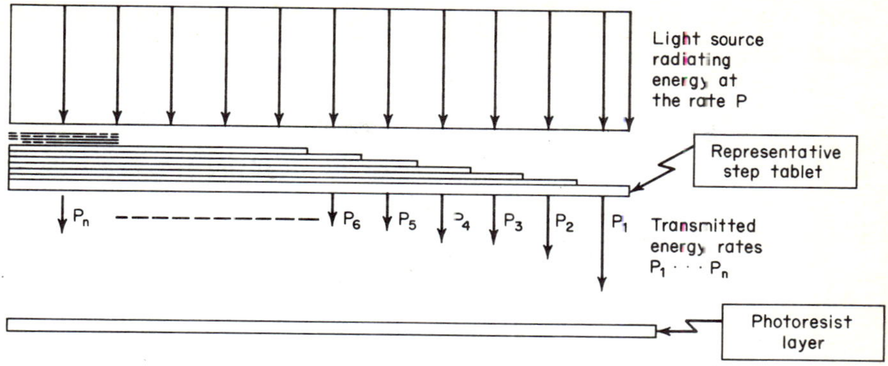

In this sketch, the changing transmission characteristics of the steps in the step tablet are represented by an increasing thickness of homogeneous layers. Radiant energy is delivered at a constant rate P from the light source uniformly over the surface of the step tablet. Each increasing step has an attenuating effect on P; it reduces the rate at which energy is transmitted. The quantity of energy Q that the photoresist beneath a given step will see is a direct function of the rate P_n at which the energy is being transmitted over the surface area of the step multiplied by the time t of exposure. The varying opacity of the steps reduces the intensity transmitted accordingly. As the time is constant for all steps, the energy that the photoresist receives will decrease with increasing opacity. A minimum quantity of energy Q_m must be reacted by a given thickness of resist for insolublization to occur. When the value of Q_m is approximately equal to the energy received by the resist beneath the sixth step Q_6, the exposure time is adequate for many types of wet-film resists. The condition of the resist at the sixth step does not indicate the condition of the image but rather serves as a visual indication of the quantity of light that the resist coating will see at the exposure time with the particular printer. Resist exposed during the test outside the step tablet is a good indication of the probable image condition.

Fig. 4-12 Step-tablet operation.

step endures processing, it should be evaluated as the threshold step. Otherwise, the last solid step should be used.

The appproach used depends upon what is known about the equipment and resist. If an approximate exposure time (or scan setting for moving-light-source printers) is known from past experiences, only an estimated adjustment is usually necessary. This approach is often used for day-to-day process control and similar instances.

The first step in determining the exposure time when the resist and/or printer are unfamiliar is the preparation of six test panels. This quantity is usually adequate for most resists in this range because the exposure time for many fixed-light-source printers is between 1 and 6 min (except for KOR). A very intense light source at close proximity is required to produce exposure times of less than 1 min. A single large panel can also be used in lieu of six smaller ones simply by covering the unexposed portions of the panel while testing. There is the consideration of heat affecting the shorter exposure times while testing the longer times with some printers, however, and single panels are preferable.

After panel preparation, an initial exposure of 1 min should be given one panel followed by developing and dyeing using the proposed methods. If there are alternative developing methods, particularly vapor degreasing, one panel should be tried for each method. After processing the panel(s) at a 1-min exposure, the condition of the resist should be noted as well as the threshold step number. If the value of the threshold step is less than 6, further exposure is indicated. One minute increments are usually satisfactory. In the event that the exposure time of 1 min produces a step greater than 6, the exposure time should be halved on the next panel to zero in on the actual time. Care should be taken to eliminate random radiation and to reduce errors. An exposure error of only 10 s which is not significant at 5 min becomes very important in a 15-s exposure. Also, the numbering system on the step tablet may start at different places; e.g., the clear end may be numbered 1, or the first step may be the first one numbered, with the clear zone disregarded altogether.

Scanning- or moving-light-source printers such as those shown in Figs. 4-9 and 4-10 require a little different approach. The exposure time is varied according to a dial setting which controls the velocity of the light source as it passes over the part. The time is usually very brief and the light sources very powerful compared with fixed-light printers. A high velocity setting gives a short exposure time, and vice versa. In this instance, the actual exposure time is not determined, but rather the dial setting to produce the desired step is. Most scanning printers have dial indicators that are numbered and evenly spaced, for example, 0 to 80 units. However, the relationship between the exposure time and dial setting is usually not linear but resembles the curve shown in Fig. 4-13.

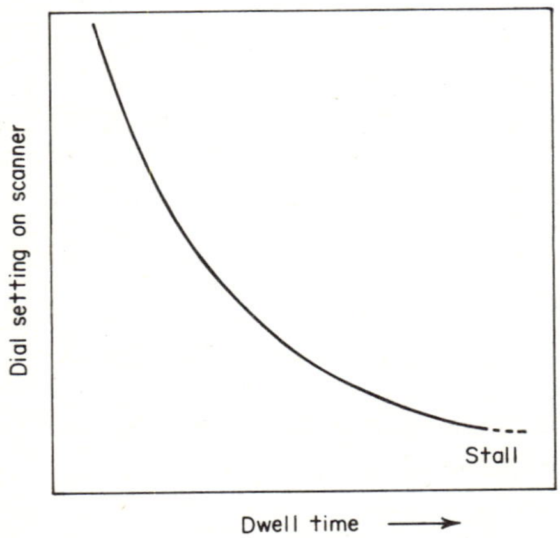

Fig. 4-13 Relationship of dial setting and exposure time for most types of scanning printers.

At one end of the scale near the setting at which the drive mechanism stalls, a slight movement of the dial may have a drastic effect upon exposure time. At the high-speed end of the scale, the same amount of movement has only a slight effect. With these factors in mind, it may be advisable to note the scanning time at three or four settings, make a simple plot, and note the shape.

To begin the actual determination, three exposures are made at 20, 50, and 80 percent of the scale. After developing and dyeing, the panels are examined and the threshold steps noted. One or two more exposures in the indicated direction will usually accurately locate the desired exposure setting. Test panels should be large enough to allow a generous border around the step tablet. The condition of the resist in these areas at the proper exposure time should be hard, glossy, and deeply colored if dyed.

Calculation Method. The calculation method is actually an extension of the step-tablet method and could have well been included there. For some of us who are not inclined toward mathematically estimating things, however, a very basic method could have been confused. It has therefore been included in a separate section.

As indicated earlier, the light transmission of each succeeding step decreases uniformly on the step tablet. For each exposure time, a proportional threshold step is achieved. If the exposure time and corresponding threshold step are known as a reference point, it is possible to estimate the exposure time to achieve the desired step using the following relationship:

$$\text{Corrected exposure time } t_c = \text{reference exposure time } t_R(\sqrt{2})^{\text{step at } t_C - \text{step at } t_R}$$

or

$$t_c = t_R(\sqrt{2})^{S_C - S_R}$$

As indicated, to use this method it is only necessary that one exposure with an identifying step be made. For example, if the reference exposure time was 4 min and yielded a number 9 step, the exposure time to reach a number 6 step would be

$$\begin{aligned} t_c &= 4(\sqrt{2})^{6-9} \\ &= 1.41 \text{ min, or } 1 \text{ min } 25 \text{ s} \end{aligned}$$

Under the same conditions, a 1-min exposure yielded a step of 5. To get a step of 6, the exposure time would be

$$\begin{aligned} t_c &= 1.0(\sqrt{2})^{6-5} \\ &= 1.41 \text{ min} \end{aligned}$$

TABLE 4-4 Exposure-Time Multiples to Obtain a Step of 6

Threshold step at reference exposure time	Exposure-time multiple K
1.0	5.7
1.5	4.8
2.0	4.0
2.5	3.4
3.0	2.8
3.5	2.4
4.0	2.0
4.5	1.7
5.0	1.4
5.5	1.2
6.0	1.0
7.0	0.71
8.0	0.50
9.0	0.36
10.0	0.25
11.0	0.18
12.0	0.13

Because 6 is the most sought after threshold step, Table 4-4 has been prepared to assist in determining the exposure time from a reference step. To use the table, a single panel is printed for a time adequate to insolubilize one or more steps. The threshold step is noted, and one refers to the table to get an exposure multiple K. An estimated exposure time can then be determined by multiplying the reference time t_R by the multiple K from the table. For example, an 8-min reference exposure yields a threshold step of 11:

$$t_c = Kt_R$$

where

$$K = 0.18 \text{ for a step of 11 (Table 4-4)}$$

$$t_R = 8.0 \text{ min}$$

or

$$t_c = (0.18)(8.0)$$

$$= 1.44 \text{ min}$$

It should be necessary to print only one panel for reference and a second panel to check the accuracy of the estimate when using this method. As with the conventional step-tablet method, the resist film at the proper exposure time should be hard and glossy.

To use the calculation method with scanning printers, it is necessary to convert the dial setting to exposure time by preparing a curve, as shown in Fig. 4-13. The time need not be the dwell time over the part in order to prepare the curve. In fact, the time to transverse the entire frame length for each sitting will do as well, because it serves only to obtain a dial setting. For example, at a dial setting of 80 the threshold step was 3. From the sample curve, 80 corresponds to 1.3 min:

$$t_c = t_R K$$

where

$$K = 2.8$$

$$t_R = 1.3 \text{ min at } 80$$

$$= (1.3)(2.8)$$

or

$$t_c = 3.64 \text{ min}$$

Returning to the time–dial setting curve, a dial setting of 34 was indicated to obtain a 3.6-min exposure time.

Inspection. Inspection is the simplest method of determining a usable exposure time, although it must be used with care or the results can be misleading. The concept, in which the condition of the exposed area is the criterion, is a simple one. Underexposed resist areas tend to be lost or damaged by the developing and dyeing operations, whereas properly or overexposed zones are hard, glossy, and lacquerlike. All that is necessary is to run a series of exposures and note the minimum time required for the criterion to be met.

There are some serious precautions that must be exercised with this method. If the time increments are too great, there is a tendency to select an overexposed zone even though it may be the first one with a glossy finish. Thin coatings on roughened surfaces may not be as easily judged as thicker coatings on smoother surfaces, and combinations thereof can cause considerable confusion. This unquantitative approach is based upon the appearance of the coating, which obviously can be influenced by numerous other factors. To use this method, enough zones must be exposed to ensure a gradual transition from complete underexposure through partial to an overexposed condition. The time increments should be short and of the same duration for best results. In addition, the surface condition of the part, the coating thickness, and the coating method should be as unvaried as practical.

On a macro basis, there is little doubt that one experienced in the photoresist art can select an adequate exposure time using this method. However, for most applications more qualitative methods should be employed.

Table 4-5 gives the results of a series of exposures. A time of 3.0 min was selected on the described results.

Methods Involving Image Accuracy. Methods involving image accuracy are the most fundamental methods of determining exposure. Two considerations are involved in these methods: accurate reproduction of the photomask and the visual condition of the resist image. There are three related methods to be considered, two of which involve actual measurement of the photoresist image and the third, a comparison of resist images of known dimensions. In order, these are (1) conventional line-width measurements, (2) step tablet–line width measurements, and (3) resolution patterns.

TABLE 4-5 Exposure Time by Inspection

Exposure time (min)	Condition of resulting image
1.0.	Lost during developing obvious under exposure
2.0.	Image endured developing but was flat and dull
3.0.	Glossy, hard with good dye retention
4.0.	Same as 3 min

Estimated exposure time between 3 and 4 min.
RESIST: KPR-3.

The conventional line-width measurement method is, as implied, a series of timed exposures in which the resist image is compared with the photomask used for all exposures. To start, a series of test panels are cleaned and coated with resist; six are usually adequate. As with any exposure test, the preexposure processing must be representative of the production methods. Starting with a photomask typical of the parts to be made, a test panel is exposed for 1 min with a fixed-light-source printer, or 80 percent of scale on a scanning printer. The parts are developed and processed through the normal postexposure cycle and examined. If underexposure is indicated, the remaining parts are processed at longer times, as indicated in other exposure methods. When the exposure time indicates that the image is satisfactory from an inspection standpoint, measurements of the line widths and photomask should be made and compared. One method of aligning points on the resist image and photomask is shown in Fig. 4-14. The accuracy of acceptable resist images is determined and an exposure time selected based on accuracy. The longest exposure time producing accurate, glossy images is preferable. How

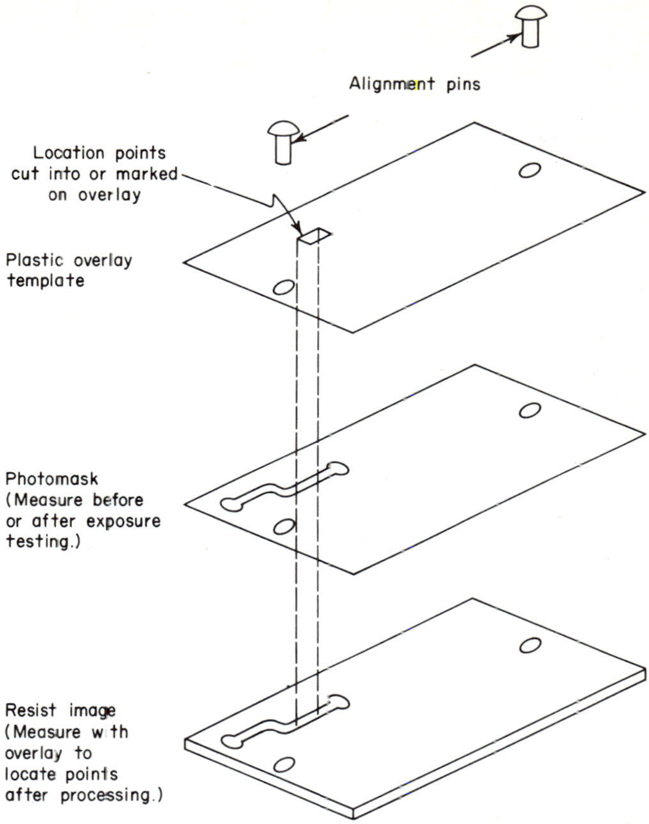

Fig. 4-14 Location and relocation of points on photomask and image for line-width measurements.

accurate is "accurate" is an obvious question that depends upon the nature of work being processed. For most macrowork, an average of ± 0.0003 in. has been satisfactory and readily maintained. This method is discussed in detail in Chap. 6, on solid-film resists, and is a primary method there.

The step tablet–line width method is somewhat similar to the conventional method except that a single line is placed under a step tablet and exposed. Line widths under each step are measured and the most accurate step selected. Exposure time is then estimated from the actual time and the percent transmission. The condition of the resist and the exposure are checked by an exposure at the calculated time. For example, a photomask (Fig. 4-15) is placed beneath a step tablet with the image emulsion against the photoresist, and it is exposed. The exposure time must be estimated, but of a duration sufficient to ensure overexposure.

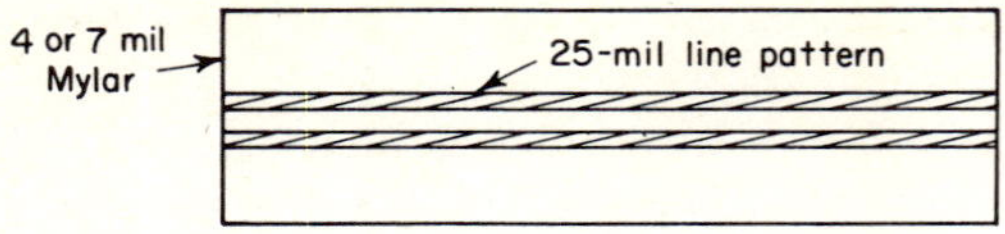

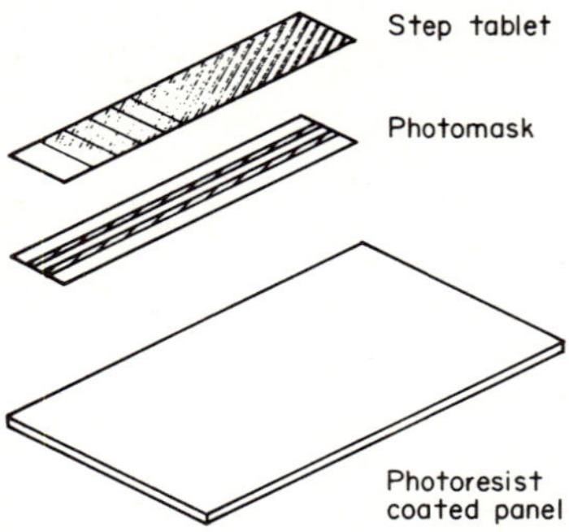

Fig. 4-15 Method of determining exposure time by exposing beneath a step tablet. After exposure and processeing, line widths in each step are measured. The step with the most accurate lines is used to estimate the exposure time.

Normally, about 6 or 8 times the point at which the resist would be insolubilized is adequate. After exposure, the part is developed and dyed (if necessary) and the resist image compared with the photomask. An equivalent exposure time is estimated from the percent transmission of the step multiplied by the actual time. Table 4-6 gives some values for the first eight steps, but the actual values are best determined by calibra-

TABLE 4-6 Results of Exposure of a Pattern Beneath a Step Tablet (Kodak No. 2 Tablet)

Step	Adjustment to original exposure time	Equivalent exposure time at 8-min original time (min)	Deviation of resist from photomask (mil)	Condition of resist image
1	0.95	7.6	+0.9	Hard, glossy
2	0.75	6.0	+0.5	Hard, glossy
3	0.43	3.4	+0.2	Hard, glossy
4	0.30	2.4	+0.1	Semigloss
5	0.23	1.8	+0.0	Flat, dull
6	0.19	1.5	−0.1	Ragged lines
7	0.16	1.3	.....	No image
8	0.14	1.1	.....	No image

Note: Step 1 was clear

tion using a densitometer. The table also gives some experimental test data as an example.

From the data in Table 4-6, the exposure time would lie between 6 and 3.4 min. By simple interpolation, an exposure time of 4.3 min could be reached, which would be quite satisfactory. A plot could also be made, which would give similar results of about 4.4 min (Fig. 4-16). An exposure check at the indicated time should be made using a typical photomask and the results noted. For control purposes, the threshold step at the indicated exposure time should be determined using a step tablet.

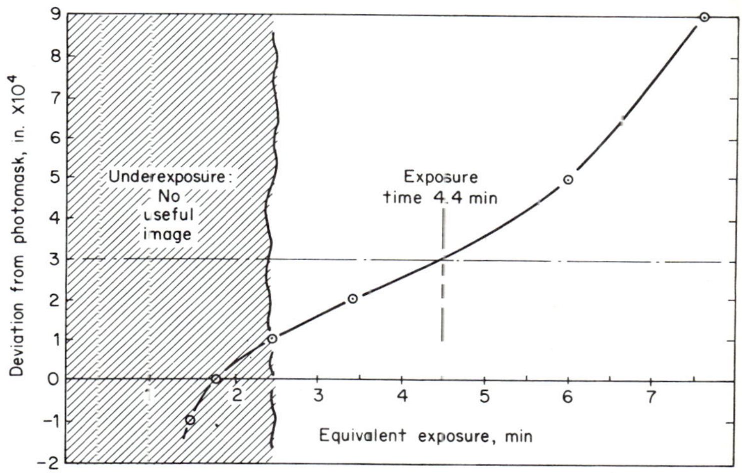

Fig. 4-16 Plot of results from pattern beneath step tablet (Table 4-6).

Resolution patterns (Fig. 4-17) contain lines and spacings decreasing in size from a pattern easily resolvable to one that is below normal resolution capabilities of the photoresists. By performing a series of exposures at varying times with a resolution pattern, the time that resolves the finest pattern commensurate with the condition of the resist gives the exposure times. Resolution patterns are particularly useful in microelectronics for determining exposure times.

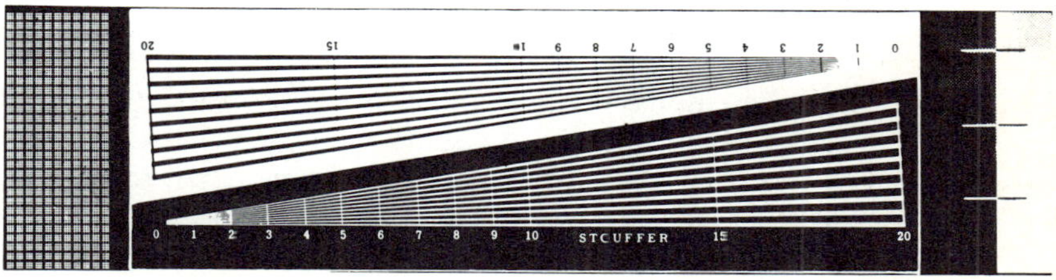

Fig. 4-17 The Stouffer Resolution Pattern. (*Stouffer Graphic Arts Equipment Co., South Bend, Ind.*)

Like the step-tablet method, a series of exposures beginning or encompassing an underexposed condition through adequate and preferably overexposure is performed. It is good practice to run a step tablet along with the resolution patterns to indicate the relative degree of exposure. When using a pattern such as the one shown in Fig. 4-17, where the lines and spaces gradually decrease, it is preferable to judge by the series of lines in the solid background. The portion of the pattern with a solid, isolated series of resist lines can narrow and lose adhesion, which gives inconclusive results. In spite of the similarity, the former pattern is more resistant to this than the latter and generally gives a more reliable indication.

When exposure, developing, and related processing steps are completed it is desirable that the patterns be etched to a depth sufficient to clearly indicate whether scumming which was not readily visible has occurred.

Comparison of Methods

Table 4-7 shows exposure times arrived at using the various methods discussed previously. As may be seen, methods involving measurements can be reasonably close. Inspection, though producing usable images, varied considerably from the other methods. This is not unexpected when the nature of the method is considered.

TABLE 4-7 Comparison of Methods

Test method	Criterion of evaluation	Indicated time (min)
1. Step-tablet	Expose to a solid threshold step of 6.	4.5
2. Calculation	Estimated from a partial step 2 at 1-min exposure	4.8
3. Inspection	First glossy finish	3.0
4. Line-width-accuracy (step-tablet)	Longest accurate exposure time	4.4

RESISTS TESTED: KPR-3 and Waycoat 20. LIGHT SOURCE: Fixed-light-source printer with 1-kW mercury-vapor lamps at 3 ft.

Control of Exposure

If one point should be emphasized in the discussion of exposure, it is the importance of day-to-day controls. Regardless of how exact the exposure-time determination has been, it is of little value if not maintained. Light sources and other factors change in even the most carefully controlled process, and checking exposure time daily is one method of preventing serious problems.

Generally the step tablet is employed for control purposes. Exposure times will not vary substantially unless something is drastically wrong, and this can usually be corrected with one or two tests. The condition of the resist as well as the threshold step should be noted. A sample of a typical circuitry pattern exposed next to the step tablet is a good indication of whether scumming and related phenomena will be likely to occur. Further, minor adjustments in developing time, the prebake time, or other cycles can be conveniently made to correct noted deficiencies before production parts are lost.

It should be pointed out that there may be no visual indications when light sources go bad. Lamps and power sources deteriorate unpredictably sometimes because of damage or differences in manufacture. Variations in line voltages from unusual demands on in-plane power can cause problems in exposure. A slight voltage drop can significantly affect the quantity of ultraviolet radiation produced by some light sources (including mercury-vapor lamps) with little visual change.

Oxygen Effects

Atmospheric oxygen has long been known to interfere with polymerization (4,5). Oxygen interferes with photopolymerization as well. The initial reaction between light and the sensitizer occurs without interference from oxygen in the exposed areas. Once free radicals are formed by the light-sensitizer reaction, polymerization can proceed. However, at this point the oxygen present in the photoresist layer can also begin to react with the free radicals, thereby reducing the number available for polymerization. Competition for the free radicals between oxygen and reactions resulting in polymerization continues during exposure as oxygen diffuses from the unexposed areas. Except in unusual instances, the amount of sensitizer is adequate to consume the available oxygen and allow adequate polymerization to occur.

When the reaction between the free radicals and oxygen predominates, the effect on the resist film is underexposure. The observable effect is loss of image quality or pattern damage. Some photoresists are affected more than others by oxygen. Resists which can be used in projection, such as KOR, are formulated to overcome the effect of oxygen. Film thickness and light sources may also be factors (6). Thicker films are better able to endure the loss of thickness and still be usable. Long exposures with low-energy light sources may be less effective than brief exposures to high-energy sources because of oxygen effects. Mathematical models for oxygen consumption in photopolymer systems have been formulated (7,8).

There are several methods of reducing or eliminating the effects of oxygen during photopolymerization. One method that is used in solid-film photoresists (Chap. 6) is to cover the surface with a transparent plastic film (9), although for microwork it is possible to process small parts in an inert-gas environment. Adequate coating thickness, prompt processing, and exposure with light sources of adequate power can usually eliminate or avoid this problem.

In addition to preventing polymerization, oxygen can promote polymerization in the presence of light. Certain kinds of dyes, such as fluorescein, in combination with a reducing agent, such as thiourea, react with light in the presence of oxygen to initiate polymerization, thereby functioning as a photosensitizer (10–12). In fact, dyes are among the most rapid and efficient sensitizers. Polymerization in the form of scumming can also be brought about by light and oxygen. Coated parts stored in the presence of oxygen can become sensitive to yellow safelights over periods of several hours and not develop cleanly after normal exposure. This phenomenon has been observed with some of the polyvinyl cinnamate resists in both white and yellow fluorescent lighting. Apparently, the exposure time varies for this phenomenon but has not been noted to occur in less than 4 h (13). Kroll (14) reported that a storage of 16 to 40 h in safelights and air could cause the phenomenon to occur after regular exposure and developing. Humidity effects have not been noted. Positive methods of controlling this phenomenon include rapid processing (<4 h) after coating and storage in dark areas or in a nitrogen-atmosphere cabinet when storage is necessary in yellow safelight conditions. (Metal-etch resists were not noted to scum under similar conditions.)

Scumming has also been attributed to smog. Engvall (15) showed that scumming of KPR could be caused specifically by ozone, one of the known oxidants in smog. Smog oxidants produced a similar scum. KMER did not scum, even in the heaviest smog, although light scum was noted with KTFR. Methods listed to eliminate this problem included processing in a nitrogen atmosphere or when smog levels are low. The use of an activated-charcoal filter for removal of the harmful oxidants was also recommended. An oxidant level of less than one part per hundred million (pphm) caused little apparent harm, whereas 5 pphm did result in scum formation.

DEVELOPING

Upon entering the developer, the exposed areas have undergone substantial physical changes due to polymerization, which has increased solvent resistance and adhesion to the substrate. Unexposed areas have not been significantly polymerized and can be redissolved in the original solvents. As most of us have observed, the developing has some effect on both the exposed and unexposed areas. Unexposed areas are severely swelled and partially dissolved by the action of the developer. The exposed image areas are much less affected, although some swelling and softening may occur.

In the unexposed areas, the polymer chains form many polymer-to-polymer contacts and tend to bunch together in groups called micelles. Such alignments are not the result of electron sharing but rather weaker or secondary attractive forces between sites on the polymer molecules. During developing, the solvent molecules break down or neutralize the intermolecular attractive forces by penetrating and attaching themselves to the polymer molecules. Individual molecules and small micelles go into solution followed by further penetration of the larger aggregates, swelling, and more dissolution. Once in solution, the larger aggregates are penetrated and further broken down. With the proper selection of solvent molecules, a solvent shell is formed around the polymer molecule, which is also attracted to other solvent molecules, and dissolution is obtained. In the image areas, the action of the developing solvent within the limits of time and temperature cannot dissolve the large polymer molecules formed by polymerization during exposure. Swelling may occur because of the penetration of aggregates of polymer molecules by the solvent molecules, however.

Proprietary developing solutions are formulated to allow removal of the unexposed areas while minimizing swelling of the exposed image areas. A fairly wide latitude of time and temperature is permissible for most applications, as indicated in Table 4-8. Developing solutions are usually mixtures of solvents. One solvent is a true solvent for the unexposed resin and the other a compatible dilutant. The latter component usually makes

TABLE 4-8 Typical Developing Parameters for Some Resists

Resist	Developing solutions and times		Some alternative solutions
KPR KPR-2, KPR-3, and KPR-4 KOR Waycoat RC and 20	KPR developer KOR developer Waycoat developer	3 to 5 min at ambient temperature	80 vol percent xylene 20 vol percent cyclo-hexanone Vapor degreaser with hot trichloroethylene
DCR- 3118 and 3140	DCR developer	30 to 60 s at ambient temperature	Xylene spray
KMER Waycoat PF CMR-5000	KMER developer Waycoat PF developer CMR developer	2 to 4 min at ambient temperature	60 vol xylene 40 vol Stoddard solvent

up much of the solution to provide a wide latitude in processing by suppression of the action of the true solvent.

The actual developing process is accomplished by immersion of the part in the solution with agitation for a minimum time. Figure 4-18 gives an illustrated description of the process. Note that some thickness is lost by the image in the sequence, which represents oxygen effects in a token sense. Some resists are affected less than others, as iterated earlier.

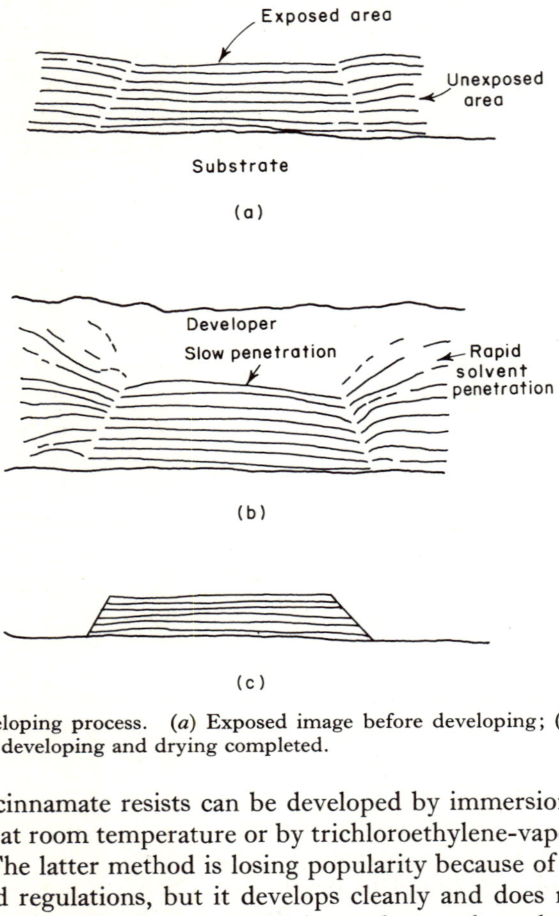

Fig. 4-18 Developing process. (*a*) Exposed image before developing; (*b*) swelling and dissolution; (*c*) developing and drying completed.

Polyvinyl cinnamate resists can be developed by immersion in proprietary solvents at room temperature or by trichloroethylene-vapor degreasers at 185°F. The latter method is losing popularity because of air-pollution problems and regulations, but it develops cleanly and does not require a final water rinse. Parts come out clean, dry, and ready for dyeing. Resists such as KPR-2, KPR-3, or KPR-4 and Waycoat RC or Waycoat 20 usually require a spray of hot solvent upon or during the immersion in the degreaser. Some care must be exercised in both the predevelopment treatment and the actual degreaser development for best results. Before degreaser development, a thorough prebake and exposure are required to fortify the image to withstand the highly active hot solvent. Prolonged exposure to the degreaser can result in excessive pinholes, loss of image definition, pattern damage, and an overall loss of image quality. Trichloroethylene is not a particularly useful developer when cold. It penetrates rapidly into both the exposed and unexposed areas, and considerable damage may result before clean developing can occur.

Developing of any of the negative resists is readily accomplished in slit tanks such as the one sketched in Fig. 4-19. Proper tank designs offer a generous volume so that frequent changes are not necessary; they have a minimum surface area to reduce evaporation losses. An inclined sump at the bottom of the tank provides a collection point for contaminants and allows easy cleaning. The most satisfactory construction material in the author's experience is welded stainless steel (for example, 304L, 316, 321, etc.). Polypropylene and aluminum have been used, however.

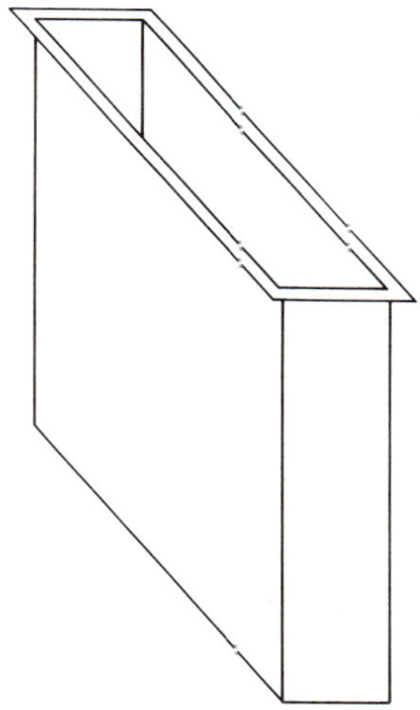

Fig. 4-19 The slit tank for use with volatile processing solutions.

Hunt and Kodak resist can be satisfactorily developed over a fairly generous time range. Good results have been obtained using a two-tank system in which the initial immersion time is 2 to 3 min followed by a 1-min immersion in tank cleaner solvent. Agitation of the solution is necessary to ensure rapid and complete developing. Solvent pumps or stirring motors with agitators are adequate. Because all the solvent developers are flammable, only explosion-proof motors should be used near the equipment, and proper care should be taken in this respect in the surrounding areas. DCR develops more rapidly than most resists, usually in a minute or less.

Solvent developers are not excessively volatile, but all contain substantial quantities of xylene. Developer tanks should be fitted with covers when not in use and adequate ventilation provided. Racks should be designed to ensure that the boards do not come in contact with any obstruction during the developing and subsequent operations. Resist coating is exceptionally sensitive to damage when swelled by the developing solvent.

Control of the developing solution is relatively easy. For many operations, simple observation on the part of the operator is all that is necessary. When the parts are not completely developed, the solvent is replaced.

Quantitative means can also be used to control the developing solvent. When the dissolved resin reaches a certain level, the developing action becomes incomplete, and a scum of resist will remain on the parts. Often, the scum will not accept dye. It can remain undetected, with poor plating adhesion or incomplete etching the eventual result. The incidence of losses of this type will depend upon the number of inspection steps and may not be of concern in some instances. However, when large numbers of boards are being continually processed, a significant number of parts could be lost before the problem is corrected. For continuous processes, the area passing through the fixed solution volumes may be noted. When the maximum area commensurate with good developing has been processed, the solutions are changed. Similarly, periodic changes of solution can be made based upon the time which the solution has been in use.

For intermittent production or when the area or time methods do not appear feasible, a third method can be used. The method simply involves determining the quantity of dissolved solids in the developer. A sample of the developing solvent is weighed, carefully brought to dryness over low heat, and reweighed. Expressed as the percent solids, it is an indication of the level to which the developer has been spent. Actual values will vary for numerous reasons. The author's experience has indicated that less than 1 percent solids in the most contaminated of two developing tanks can cause problems in both plating and etching processes with some polyvinyl cinnamate resists. Further, the requirements for plating are usually more stringent than for etching. Regardless of which type of resist is used, it is normally possible to establish a controllable solid level for the developing solvent.

Spray developing with solvents at ambient temperature is very effective. Developing is both quick and thorough if properly performed. A design specification is available from Kodak.

DYEING

Dyeing is a convenient method to render the resist image more visible for examination and inspection. Upon entering the dye solution, the part has usually been transferred directly from developing and is therefore soft and swollen. The solvents in the dye solution are similar to those in the developing solution, and there is no incompatibility from drag-in. Dye is absorbed by the resist surface rapidly upon immersion, and only about 20 s is required in most instances. Some loss of chemical resistance usually occurs when dyeing, but the increased image visibility is frequently necessary. If it is possible to eliminate the need for inspection, dyeing should be eliminated. Dye solutions are expensive to procure and messy to use.

Some solutions combine both dyeing and developing. Although they eliminate a processing step, the cost should be considered. Dye solutions cost several times more than the developer, which loads up with solids and must be dumped. The dye normally has quite a long life and is dumped much less frequently. Consequently, the life of the combination solution is predicated upon the least expensive constituent.

The actual operation is usually carried out in tanks similar to those used for developing. Stainless steel is preferable to avoid contamination.

WATER RINSES

In the processing of photoresists, only one or two water rinses are normally required. A spray of moderate force is required to remove the exposed portions of the pattern after immersion developing and dyeing. High-pressure sprays can cause damage to the exposed areas. Temperature is not a critical factor, but excessively hot (>80°F) or cold (<60°F) water should be avoided. There is little evidence to indicate that high-purity water is required for this rinse. Tap or industrial water is often used.

TOUCH-UP AND SCANNING

It is possible with careful process control to avoid touch-up and scanning. Not infrequently, more time is consumed by scanning and touch-up than by stripping and reapplying a new pattern. Damage from the additional handling can also reduce yields. Conversely, these steps act as a review of the process, a point at which defects can be located and traced to their source, e.g., photomask effects or contaminants. Moreover, process losses from continuing defects are avoided or reduced before they reach substantial levels. The cost and complexity of the parts, their susceptibility to small defects, and the level of process control maintained are major factors in deciding whether to touch up.

Touch-up consists of fortifying the resist pattern and/or removing extraneous resist from the nonimage areas. Defects in the resist image are usually touched up with a nonphotosensitive resin that is resistant to the processing solutions. In addition, the touch-up resin should be dissolved in solvents which do not damage the resist coating and must be adherent to the photoresist resin. It is desirable that the resin be of contrasting color and have good pot life. Inks (vinyls, acrylics) used for screen stenciling are normally useful as touch-up inks.

POSTBAKING

Postexposure baking promotes thermal polymerization, which in turn increases chemical resistance and adhesion. Resist films are embrittled to varying degrees in the process. In many instances postbaking may not be necessary, and a simple air dry to expel absorbed developing solvents may be adequate. Modern resists can adequately withstand most acid etchants with a 20-min ambient air dry. Plating processes invariably place greater demands upon the resist image than acid etchants, and postbaking is usually necessary. Caustic chemical milling solutions also require excellent resist adhesion. Prebake cycles can often be used.

STRIPPING NEGATIVE RESISTS

After the photoresist has performed its function, it is necessary to completely remove it from parts. Incomplete removal from circuit boards can cause serious problems in subsequent processing. Plated patterns will be difficult to etch completely because of residual resist, and lamination problems may occur with etched multilayer parts.

Prior processing determines the ease with which photoresist can be stripped. The extent of the postbaking is quite significant in this respect and, as indicated in that section, should be held to a minimum commensurate with acceptable yields for the intended use. Good stripping solutions that aid in stripping are readily available commercially. The primary job of strippers is, of course, to remove the resist without damaging the substrate material excessively. Two basic types of solutions are used to remove photoresists: (1) cold-solvent strippers and (2) hot strippers.

Cold-solvent strippers are the most common type. Their primary constituent is methylene chloride fortified with lower alcohols and ketones, an acid, and a wetting agent. Other constituents may include paraffin wax to retard evaporation, phenols, cresols, xylene and hydrocarbon solvents, dioxane, aromatic aldehydes, and glycol ethers and esters. The stripping action involves swelling and lifting of the resist film with some dissolution. Scrubbing action is usually necessary for complete removal of the residues. A typical process is shown in Table 4-9 for this type of stripper. The process times given are based on a solution that will not excessively damage the substrate within 3 min but will remove at least 70 percent of an exposed area covered with resist within 1 min by soaking. Various tests can be devised for evaluation of stripping solutions of this type, but the effectiveness of the stripper in removing the photoresist must

TABLE 4-9 Example of Hand Photoresist Stripping Operation

Operation	Time
1. Soak parts in stripping solution	30 s
2. Scrub, soften resist with soft brush. Keep wet with solution	30 s
3. Rinse briefly with fresh stripping solution to remove loosely adhering resist	30 s
4. Water-rinse to remove residual solvents	1 min
5. Dry	As required

be weighed against the damage done to the substrate materials. Usually, a safety factor of 2 is adequate. That is, the time of resist removal must only be about half of the stripping time.

The second type of stripper utilizes a hot-solvent solution that removes resists primarily by dissolution. Both alkaline and acid solutions are available. Phenols, cresols, and related derivations are utilized in both acid and alkaline solutions but are not a desirable constituent because of contamination of parts. Acid solutions may contain chlorinated benzenes, high-boiling aliphatic chlorinated hydrocarbons, mineral acids, aromatic aldehydes, and other compounds. Use of these strippers is generally limited to metal parts or parts with no exposed plastics on the substrate because of the damage that may occur. Recent formulations show less tendency toward attack of circuit-board material and may be used to strip both plated and etched patterns (see Chap. 6).

Machines for stripping photoresist are available for rapid stripping operations. Parts are placed on conveyers and sprayed both top and bottom with stripper followed by rinsing with solvent. A final water rinse and drying complete the operation. All types of resist including positive and dry-film resist can be stripped in these machines. The Du Pont stripper is equipped with a still and used with methylene chloride. It can continuously strip and recycle the solvent (see Chap. 6). Methylene chloride will evaporate too rapidly without a still or recovery system to be sprayed.

CONTAMINATION LEVELS

After peaking the process to produce highly accurate images, contamination becomes an even more serious consideration. Dust and airborne contamination are particularly important when one uses negative-acting resists because they tend to be resolved as voids, pinholes, or thin spots in the resist image. This results in bridging and shorting between circuitry during subsequent plating, or voids, pits, nicks, and lost circuit lines when etching. At the least, greater touch-up will be required and losses increased.

As may be noted in Table 4-10, clean resist, solvents, and ambient air can eliminate many contamination problems. Resist filtration is one important control. Coating methods, such as dipping or roller coating, in which the resist is continually exposed to the parts tend to build up particulate contaminants. Spraying or whirling methods in which the resist is used only once do not tend to have this problem. In the former instance, continuous filtration through a nominal 10-μm filter of solvent-resistant materials is necessary during use. In the latter consideration, resist filtration just prior to application is necessary. Normal filtration will not remove water from the resist. The best method of control is prevention. Excessive moisture causes precipitates in the resist.

TABLE 4-10 Some Sources of Contamination

Operation	Typical contaminant
1. Resist application	Water, abrasives from cleaning, fingerprints from handling after cleaning, matter from unfiltered resist, fibers and particles of copper from board previously processed (roller and dip coating)
2. Drying and prebaking	Airborne particles, fingerprints, spurious light, moisture, smog
3. Exposure	Airborne particles, fingerprints, dried resist on vacuum frame and photomask
4. Developing	Water in developer, excess resist in developer, tacky resist particles
5. Dyeing	Excess resist, tacky particles of resist, water
6. Rinsing, alcohol dip	Insoluble particles, metal salts

Airborne contamination can be a serious problem and was so in the early days of photoresist. Some resists tend to pinhole with airborne particulate matter. An electrostatic repulsion between the resist and the particle causes a thin area which can be penetrated by etchants. Other resists are less affected with less repulsive forces existing between the resin and the particulate. This type of contamination is controllable and has been the subject of numerous studies. For macrowork, Class 160,000 or less as measured in Fed Std 209 is usually adequate (16).

One important source of contamination which invariably causes problems if not checked periodically is the vacuum frame of the printer. Airborne contamination and particulate matter can collect on the glass or plastic overlay. Photoresist residues adhere to the printing frame and are good absorbers of ultraviolet light regardless of the color or visual appearances. The photomask which is in contact with the resist surface can become contaminated in the transparent areas. Printing frames should be inspected and cleaned on a daily basis. Periodic inspection or replacement of the photomasks is necessary not only from the standpoint of contamination but for damage from usage. Most macrophotomasks have bases of polyethylene terephthalate and can safely be cleaned with water or certain solvent cleaners, such as Freon solvent. The clear plastic overlays on printing frames are of the same material. Particulate matter, such as dust, can readily be removed from printing frames and photomasks by the use of static eliminators, which neutralize the electrostatic charge on the particles for ease of removal. The devices are available as brushes or slit cleaners for photomasks.

REFERENCES

1. "Data Sheet for KPR-4 Photoresist," Eastman Kodak Company, Rochester, N.Y.
2. "Data Sheet on Waycoat RC Resist," P. A. Hunt Company, Palisades Park, N.J.
3. "Incidental Intelligence about Kodak Resists," Eastman Kodak Company, Rochester, N.Y., vol. 6, no. 2, 1968.
4. H. S. Taylor and A. A. Vernon, *J. Am. Chem. Soc.*, vol. 53, p. 2529, 1931.
5. C. E. Barnes, *J. Am. Chem.. Soc.*, vol. 67, p. 217, 1945.
6. W. M. Moreau, *Tech. Pap., Reg. Tech. Conf. Photopolymers, Mid-Hudson Sec., Soc. Plast. Eng.*, 1970, p. 137.
7. D. K. Smith, *Photographic Science Eng.*, vol. 12, no. 5, pp. 263–266, 1968.
8. J. S. Chong, *J. Applied Polymer Science*, vol. 13, pp. 241–247, 1969.
9. G. Oster, *Nature*, vol. 173, pp. 300–301, 1954.
10. S. Chaberek and R. J. Allen, *J. Phys. Chem.*, vol. 69, pp. 647–656, 1965.
11. J. D. Margerum, L. J. Miller, and J. B. Rust, *Photographic Science Eng.*, vol. 12, pp. 177–184, 1968.
12. G. Oster, U.S. Patent 2,850,445 (1958).
13. Author's data (unpublished).
14. H. Kroll, *Proc. IPC Fall Meeting Photoprinting Photoresist*, 1968, p. C-16.
15. A. E. Engvall, *Proc. Kodak Photoresist Semin., Kodak Publ. P-209*, 1969, pp. 12–17.
16. "Clean Room and Work Station Requirements, Control Enviroment," *Airborne Contamination*, Fed Std 209.

CHAPTER FIVE

Positive Resists

In this chapter, processing details for positive-acting wet-film resists will be discussed. The composition information on these resists, where available, was discussed in Chap. 2 and their response to light shown in Fig. 1-2. Compared with negative resists, the positives are comparatively new on the electronics scene. Like negative resists, they were formulated originally for lithographic plates, although the sensitizing reaction had been known for many years (1). In the early 1960s, positive resists were introduced to the electronics industry by the Shipley Company and, thereafter, others followed, including GAF, Kodak and Dynachem (Table 5-1).

Positive photoresists produce very accurate images that have moderate chemical resistance with minimal processing techniques. In general, these resists are very versatile and appear to be enduring the recent, strong industrial trend toward dry-film photoresists. Perhaps it is because of the similarity in processing steps under some conditions. For example, Table 5-2 shows a typical processing sequence. If steps 1 to 3 are contained in a roller coater with a conveyerized infrared dryer and steps 5 to 7 in commercially available conveyerized spray equipment, the process is

TABLE 5-1 Some Positive Resists Used in the Electronics Industry

Resist	Use	Viscosity (cP)	Solids content (%)
Shipley AZ-111 or AZ-111B	General etching and plating	35.7	19.5
Shipley AZ-119 or AZ-119B	General etching and plating	167.0	26.0
Shipley AZ-1350H or AZ-1350J	Microelectronics, general purpose	25.0	30.0
Shipley AZ-340	General etching and plating	11.0	26.0
Shipley AZ-345	General etching and plating	19.2	19.5
Dynachem PRS-175	General etching and plating	17.5	17.5
GAF Microline PR-102	Microelectronics, general purpose	1.8	16.6
Kodak KAR-3	General etching and plating	12.0	26.5

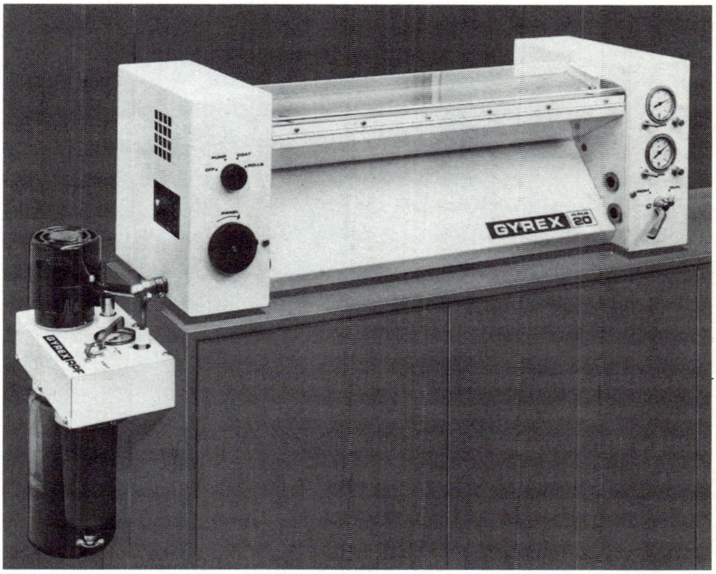

Fig. 5-1 The Gyrex 20 double-sided roller coater. *(Gyrex Corp., Santa Barbara, Calif.)*

TABLE 5-2 Processing Sequence for Positive Resists

Step	Comment
1. Apply resist	Roller-coat, spray, or dip
2. Air dry	Expel solvents
3. Prebake	Expel solvents, harden film, improve adhesion
4. Expose	Define image
5. Develop	Delineate image
6. Rinse	Wash off resist residues
7. Dry	Remove water from part
8. Examine and touch up	Correct image defects
9. Postbake	Further harden coating and improve adhesion

reduced to the three essential steps encountered in dry-film processing. Dry films, however, simultaneously coat both sides of the part with resist, which is not possible with most roller coaters. However, double-sided roller coaters are now commercially available (Fig. 5-1), but drying is still not readily conveyerized for boards wet-coated on both sides.

RESIST APPLICATION

Coating thickness is not as critical for processing positive resists as with negative resists but is nevertheless very important. All subsequent processing steps can accommodate thickness variations with certain limitations and still produce usable images. If the coating-thickness variations exceed the allowable limitations, processing problems will be encountered. With positive resists, the exposure and coating thickness must be particularly well coordinated. For example, when the thickness of resist exceeds the exposure range, scumming or no image at all may result. If the film is too thin, failure may occur in the plating or etching operations for which the image has been prepared.

The thickness and uniformity of positive-resist coatings will, of course, vary with the coating method. Some resists are formulated with very low viscosities and form only thin coatings (Table 5-1). Generally, however, the positives should be at least 100 μin. thick for etching applications. When pattern plating is necessary, 200 to 300 μin. will usually give adequate chemical resistance. Deep etching or etching with etchants that tend to degrade resists, such as chromic-sulfuric acid mixtures, will also probably require 200 μin. or more for best results. Coating-thickness variations (uniformity) of 50 percent or more usually do not cause significant processing problems. Uniformity of the resist coating is defined as the minimum thickness in the useful area of the part over the maximum thickness in the same area:

$$\text{Uniformity} = \frac{T_{\text{min}}}{T_{\text{max}}} \times 100$$

Thickness measurements may be made in a number of ways. Examples include precision profilometers or backscatter devices, such as the Microderm thickness tester.

In the discussions of individual coating methods, thickness and uniformity will be considered for the following methods: (1) roller coating, (2) dip coating, (3) spraying, and (4) other methods. It must be assumed that clean, dry parts arrive at the coating station. Cleaning methods are discussed in Chap. 3.

As all methods may require some form of thinning for viscosity control, a brief discussion will be given on thinning effects and film formation.

Thinning and Film Formation

Although the exposure parameters of positive resists are less sensitive to coating-thickness variations than most negative resists (2), it is still necessary and desirable to obtain uniform and repeatable coatings for predictable performance. To do this, the viscosity of the photoresist solution must be adjusted with solvents at some time during its useful life. Solvents, herein, refer to a group of chemicals that are liquid at room temperature and are used to make solids or viscous liquids fluid enough for practical use. Solvent molecules do not react chemically with the materials being dissolved but are bound to the molecules of the material by secondary forces. When the coating solution is applied to the substrate, the solvent molecules evaporate, and a dry, coherent film is formed. The process is physical in nature in contrast to conventional paints, which form chemical films through reactions during drying.

Under ideal conditions and for maximum viscosity each resist constituent would be ideally solvated. To do this, the solvent molecules must overcome the attraction that individual molecules of the same type have for each other in solution and form an envelope of solvent molecules around them. If a poor solvent is used, the result will essentially be to drive molecules of similar types together in clusters, because the attraction of the solvent molecules is not sufficient to overcome the attraction that the material molecules have for each other. To be sure, many other factors must be considered, such as the molecular weight of the material being solvated and the affinity of different solvent molecules for each other, but the addition of a poor solvent to a resist solution will generally cause an unusually large drop in viscosity. Higher-molecular-weight molecules will generally be precipitated first by the action of the poor solvent before lower-molecular-weight molecules of the same kind. The poor solvent may also tend to dissolve one constituent preferentially. In other words, the poor solvent may cause drastic changes in the physical properties of the film even after it has evaporated and certainly affect the properties of the coating solution. The result is generally uneven and defective coatings.

Most photoresists contain more than one solvent, which may be categorized as follows:

1. *True solvents.* As the name implies, these are actually solvents for the resin and sensitizer. These are usually the least volatile of the solvent group and are the last to evaporate. One of the more common examples of this type solvent is ethylene glycol monoethyl ether (Cellosolve) or ethylene glycol monoethyl ether acetate (Cellosolve acetate) for the ester sensitizers. The true solvent makes up about 80 percent of the solvent system for many resists.

2. *Dilutants.* Positive resists have large quantities of high-molecular-weight sensitizer present, which may tend to crystallize from solution with improper dilution. Dilutants are generally not solvents for the resist but can be added in limited amounts to the true solvent mixtures containing the resist constituents. Examples include aromatic hydrocarbons, such as xylene, and esters, such as butyl acetate.

3. *Thinners.* With positive resists, mixtures of the true solvent and dilutants similar to that contained in the resist formulations are generally used for viscosity control. This mixture can be added to the photoresist in any proportion without causing precipitation.

Photoresist formulations contain a true solvent and dilutants. Table 5-3 lists some of the positive resists and commercial thinners. Of particular importance is the AZ-119 roller-coating resist, where the use of an improper thinner would cause drastic changes in viscosity.

TABLE 5-3 Positive Resists and Thinning Solvents

Resist	Application method	Viscosity-control solvent
Shipley AZ-111	Dipping or spraying	AZ thinner
Shipley AZ-340	Dipping	AZ thinner
Shipley AZ-345	Dipping	AZ thinner
Shipley AZ-119	Roller coating	AZ-119 thinner
Dynachem PRS-175	Dipping	PRS thinner
Kodak KAR-3	Dipping	KAR thinner
GAF Microline	Spraying	Thinner T-006

Roller Coating

Positive resists are generally more expensive than negative resists, and the economy afforded by roller coating is particularly of interest. Moreover, little or no photoresist reaches inside the plated through holes, which is very advantageous for plating operations since the resist will require exposure before removal by developing. Because exposure inside plated through holes is difficult, it is likely that resist deposited there will remain until stripping and cause pattern plating defects. This fact should be noted by those switching from negative to positive roller-coated resists. In the former instance, resist in the plated through holes retains solubility in the developer; so some flowing would be of little consequence and probably go unnoticed.

Conventional smooth rubber rollers give thin coatings with positive resists because of their low viscosity. It was therefore necessary to develop a method to roller-coat positive resists. Such a method was patented (3) by the Shipley Company; it utilizes threaded coating rollers. Each thread functioned as tiny reservoirs that impart resist to the part when deformed against the part during coating. Figure 5-2*a* shows the basic coating mechanisms that have been evolved from the patent along with the early smooth roller. The three categories are:

Type I	Smooth applicator roller and rotating doctor roller
Type II	Threaded applicator roller and stationary doctor roller
Type III............	Threaded applicator roller and a doctor blade

The smooth coating roller, discussed in Chap. 4, must be used with a rotating doctor roller. Threaded rollers come in varying thread angles and number of threads per inch. The larger the number of threads per inch, the thinner the coating for a given thread angle. Figure 5-2*b* illustrates the reason. Reducing the thread angle will increase the thickness slightly if the number of threads per inch is not changed, as shown in Fig. 5-2*c*. The thread is cut in the roller in the form of a continuous helix from one end to the other. A shoulder is left at either end of the applicator roller to support the doctor blade (or roller). Otherwise, the doctor blade would compress direct against the threads, and the pressure adjustment would become delicate indeed. It further serves to maintain the applicator roller and the doctor blade parallel for even coating thicknesses.

With the part velocity, doctor-roll (or blade) pressure, and resist viscosity constant with the proper interference setting, the thread count (threads per inch) and the thread angle will determine coating thickness. Wet-film thickness and, subsequently, dry-film thickness are determined by the amount or volume of resist transferred contained in the threads of the coating roller. Anything which increases this volume will allow a thicker deposit. This statement is, of course, subject to practical limitations,

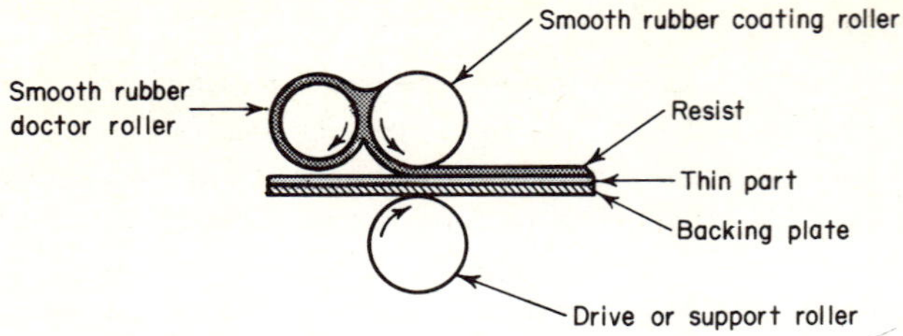

Type I. Doctor and applicator roller both rotate. The applicator (or coating) roller is smooth.

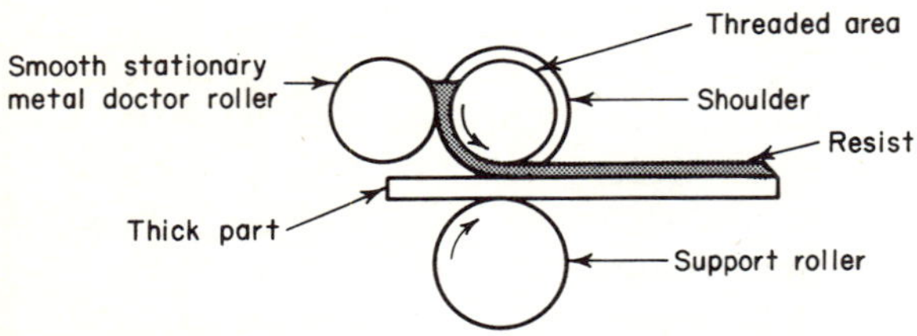

Type II. Stationary doctor roller and threaded coating roller.

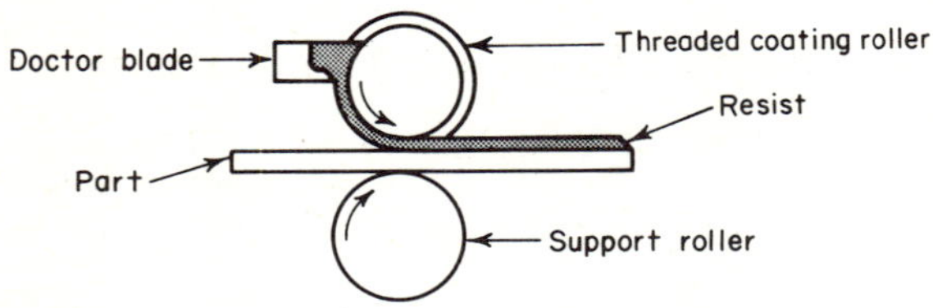

Type III. A stationary doctor blade with a slotted trough forms a dam against the rotating applicator roller. Used with micro type roller coaters.

Fig. 5-2a Basic types of roller-coater mechanisms.

in that an extremely coarse roller would deposit an uneven coating, whereas the extremely fine thread would be unable to contain very much resist and would tend to catch or scrape against the doctor roller or blade.

In operation, the photoresist is pumped from a sump to the roller, where it flows outward toward the end of the roller and is recirculated. When in operation the coating-roll grooves fill with resist, which is flowed on the part when the threads deform against the surface. After being applied, the resist coating levels and flows while still fluid to form an even coating. If the support roll is too hard, the resist may be forced into the holes by excessive deformation of the threads. For the same reason, the interference setting between the part and the coating roll is also important. Low pressure between the doctor roll and the coating roll may allow excessive amounts of resist to build up on the coating roll, and, again, resist

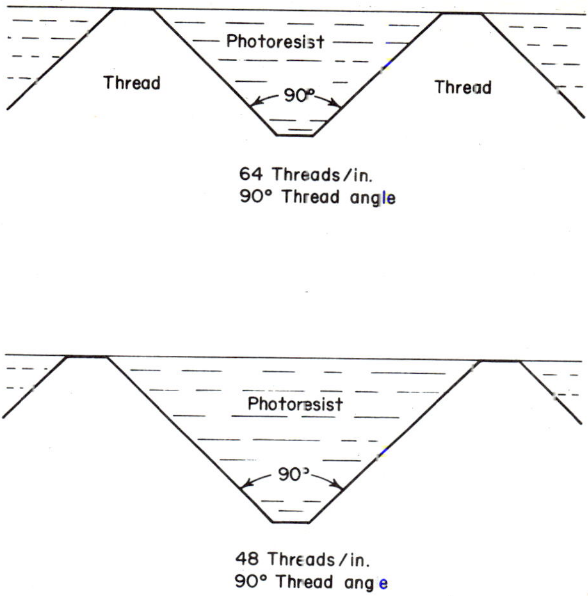

Sketch showing a cross section of threads in the applicator roller (see Fig. 5-2*a*). Increasing the number of threads per inch will decrease the volume of resist retained in the roller, and thereby the coating thickness if the thread angle remains unchanged. In this instance decreasing the thread count from 64 to 48 will increase the volume by about 25 percent and the coating thickness by about 15 to 20 percent.

Fig. 5-2*b* The effect of thread count per inch on the coating thickness.

may flow into the holes or even wet the driver roller. Very small holes may be bridged with a wet film of resist during operation, which may cause some resist to flow inside the holes upon drying. In general, the use of lower-viscosity resist will usually alleviate the problem, but the lower-viscosity resist will result in a thinner coating, and double wet coats may be necessary.

Several factors affect the average coating thickness of the resist to varying degrees using the previously described coating system. These factors are (1) resist viscosity, (2) conveyer speed, and (3) doctor-roll pressure.

Resist viscosity has s significant effect on the coating thickness and uniformity. Best results are obtained between 130 and 165 cP. At low (<110 cP) viscosity, excessive flowing will occur, and coatings will be too thin to be useful. Moreover, the resist will invariably flow into open holes and require stripping and recoating or, at least, a double wet pass for adequate thickness. At very high viscosities (>190 cP), the resist does not flow adequately before drying. Uneven coatings and poor wetting

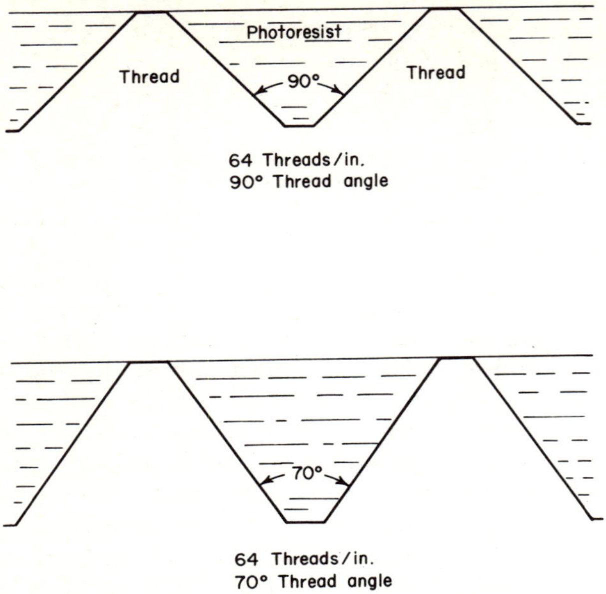

For a given pitch, the smaller thread angle (70°) will hold about 1½ times the volume of resist that the 90° thread will, but the resist deposit will be about 5 or 10 percent greater.

Fig. 5-2c The effect of thread angle on coating thickness.

will occur. With available thread angles (70 and 90°) and thread count (48 and 64 threads per inch), coatings 200 to 300 μin. thick can be attained in one pass. Coating factors of 70 to 80 percent are readily attained.

The viscosity is conveniently controlled by the General Electric #2 Zahn cup during operation, which has a recommended range of 30 to 170 cP. A range of 60 to 85 s is usually adequate for most purposes. Changing the #2 cup viscosity by 10 s may change the coating thickness by 20 μin. within this range. One precaution must be noted when using the cups. The temperature of the resist should be nearly the same each time to reduce errors. A stainless-steel dial thermometer inserted in the sump with a range of ± 1°F is adequate. The resist at an equilibrium temperature will probably be slightly above ambient temperature because of the work done on the fluid by the circulating pump. Before significant additions of thinner are made, the coating system should be allowed to run or "warm up." Viscosity is measured with Zahn cups relative to a fixed standard on the same fluid measured by more absolute means, such as the Brookfield or the Cannon-Fenske viscometer. Each of the five Zahn cups has the same fluid capacity (44 cm^3) but has a different orifice size, varying from 0.078 in. for the #1 cup to 0.208 in. for the #5 cup. Obviously, the larger the orifice, the more viscous the liquid to be measured. The #2 cup, which is most useful here, has an orifice size of 0.108 in.

To use the Zahn cup, it is totally immersed in the solution for a short while to wet thoroughly and reach temperature equilibrium with the resist. Some prefer to leave the cup permanently immersed in the sump, which does not appear to cause problems unless the resist dries on the cup and reduces the orifice size. Most sumps are thoroughly agitated by the equipment pump, but some agitation with the cup is advisable to ensure repeatable results. Timing is started when the rim of the cup breaks the surface of the liquid and ends when there is a sharp break in the fluid flowing from the orifice. After the distinctive break in fluid flow, there may be some dripping from the orifice which should not be counted. Because the test is simple and quick, it is advisable to run two in succession. If the times are within 2 s of each other and within the control range, viscosity adjustments can then be made accurately.

In regard to additions of thinner, care should be exercised not to add excessive amounts, which would require, in turn, that more resist would have to be added to bring the viscosity back within range. Because the volume of the sumps is limited to about 1 gal in most instances and the resist is expensive, it is advisable to dilute carefully to within the desired viscosity range with one addition. A rule of thumb to avoid excessive dilution when adding the manufacturer's recommended thinner is to never exceed 10 percent of sump volume with any addition (about 380 ml/gal), even if the parts show missed areas in the coating. A simple schedule can readily be worked out relating Zahn cup seconds to thinner additions and placed close to the sump as a reference. After the addition, a 3- to 5-min wait for mixing is good practice before one makes the next viscosity measurement.

Conveyer speed does not affect coating thickness with the threaded-roller system nearly as much as with the smooth roller used on negative resists. Conveyer speeds of 4 to 8 ft/min are satisfactory for most applications with the Gyrex roller coater. Doubling the speed affects the resist thickness only about 10 or 15 percent overall with thicker coatings being applied with slower speeds, which is the opposite of the effect noted with smooth rubber rollers and negative resist.

The doctor-roll pressure reduces the coating thickness somewhat as it increases. When setting up the roller coater, the doctor-roll pressure may be used to adjust the coating thickness to some extent for Types II and III coaters. Good results can frequently be obtained between 4 and 8 psig, but adjustments are best made based upon the thickness of the resist coating. The appearance of the resist pattern on the coating roller can also indicate when the proper setting has been achieved once it has been related to the coating thickness. Type I coaters are used at higher doctor-roller pressures and have little effect on coating thickness above a pressure of about 8 psig.

The interference setting is defined herein as the distance between the total thickness of the part and its support (if any), and the distance between the coating roll and the drive or support roll. In another sense, it is the total dimensional compression which the coating part and support roll undergo when the part passes between them. For systems such as the Gyrex 630, an interference distance of 0.030 to 0.040 in. gives good results and fairly constant thicknesses. Too small an interference will cause a significant decrease in coating uniformity, and too great a distance will cause excessive distortion of the coating roll and deposit resist in the holes as well as contribute unnecessary wear on the rollers.

When parts are to be plated, it may be desirable to increase the thickness of the resist by double wet coating. The parts are simply picked up immediately after being coated, while still wet, and returned through the roller coater. It is good practice to rotate the parts 90° to cover defects that may have been shadowed by obstructions during the first pass. A thickness increase of about 30 percent will result for AZ-119 in many instances. While double passes may improve the yield through reduced defects, the additional processing costs should be considered.

Double-sided production roller coaters, such as the Gyrex 20, are also available, but information regarding its operation is somewhat limited (Fig. 5-1).

When coating very thin parts that are close to the interference setting in thickness, it becomes necessary to use a backup plate to increase the thickness and rigidity of the part. Otherwise, the coating roller may contaminate the support roller with resist, which may in turn deposit a small amount of resists on parts being coated. Backup plate material should be chosen carefully, as indicated in the following comments:

1. The plate should be resistant to cleaning solvents and the resist. It should not deteriorate in such a manner as to contaminate or impregnate the dry- or wet-resist films.
2. Nonabsorbent material should be used to avoid retention of resist and cleaning solvents, whenever possible.
3. Surfaces should be flat and very smooth to accommodate resist-coated parts without damage to the resist film.
4. The plate surface should be rigid and heat-resistant.
5. Dimensions should be approximately the same as the parts being processed to avoid part deformation or applying resist to the plate. The thickness should represent that of the bulk of the rigid parts normally processed, for example, 0.062 in. thick. Thin parts can then be pro-

cessed along with the standard parts without changing the interference settings.

During use, periodic inspection and cleaning of the backup plate is necessary to remove contamination that builds up during use.

Dip Coating

With few exceptions, most of the photoresists shown in Table 5-1 were apparently formulated to be applied by dip coating if one may judge by the product data sheets which have been summarized in Table 5-3. Dipping is an excellent application method within certain limits. It is simple and rapid, simultaneously coats both sides of the part, and does not require elaborate equipment or the skilled operators used in spraying. However, resist losses are fairly high, depending upon the draining period that is allowed. Plated through-hole parts cannot be pattern-plated in most instances. When a positive resist is applied inside a plated through hole, it is difficult to remove by exposure and developing, particularly if the part is of conventional thickness (0.062 in.). The reason, of course, is that the exposure light passes through the holes in a direction parallel to the hole wall and not enough incident radiation strikes the wall coating to render it soluble to the developer. Hence, dip coating is used primarily on print-and-etch work where there is no pattern plating to contend with. An exception to this is the Shipley CTH system, which was used to produce print-and-etch parts before the advent of the dry-film resists (Chap. 6). AZ-340 or AZ-345 resists tended to form a doughnut-shaped buildup around the periphery of the plated through hole. This is in contrast to most resists, which tend to draw away from the rim of the hole and leave only a thin film, which is an unreliable etch resist. The doughnut was able to protect the parts during etching sufficiently to allow the production of parts without benefit of dissimilar metal plating.

Coating thickness varies as the viscosity of the resist and the vertical withdrawal rate from the resist. For a given resist formulation, higher viscosities will give thicker coatings, although the resists shown in Table 5-3 can be used with good results at the viscosity received from the supplier. Viscosity control can be maintained for some of the resists with the Zahn #1 cup, which has a recommended limit of 85 cP. A more accurate means may be necessary if more precise coating thickness control is required. There is considerable viscosity variation with only a small time change below 20 cP, which is the recommended lower limit for the #1 cup. The Cannon-Fenske viscometer (ASTM D445-61) is a rapid, accurate method that requires only small resist samples and would be useful on a daily basis. AZ-111 is well within the Zahn #1 cup range, which could probably be used satisfactorily for that resist.

With constant viscosity and flat parts, the coating thickness will vary as the withdrawal rate from the solution. Increasing the withdrawal rate increases the coating thickness within certain limits. Obviously, the relationship between thickness and withdrawal rate is not linear or else a withdrawal rate of 50 in/min would put on a coating several mils thick with a resist viscosity of less than 20 cP. Thickness does increase with withdrawal rate up to a certain point and, thereafter, very little, as pictured in Fig. 5-3. Zone I of the figure, where there is a near-linear variation of

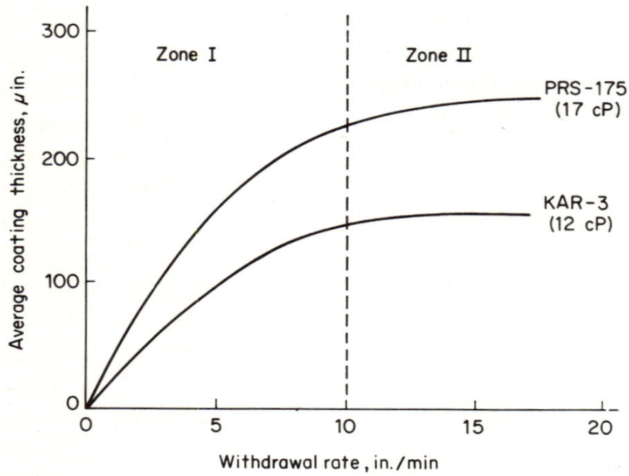

Fig. 5-3 Effect of coating thickness on withdrawal rate when dip coating.

coating thickness with withdrawal rate, is where most of the available mechanical dip coaters operate. Although Zone 1 operations are slower, coating uniformity is generally better, about 60 percent uniformity being fairly common. Zone II usually begins around 10 to 15 in/min and can produce fairly uniform coatings that are readily able to be processed. Very viscous photoresists, such as AZ-119, were not formulated for dip coating but may possibly be dipped if the withdrawal rate is very slow (<3 in/min) with a uniformity of 50 percent or more.

Spraying

Spraying is useful for coating large parts or large areas that cannot be readily coated by other means. Boards with plated through holes can be processed by spray coating with positive resists. Conversely, spraying is highly operator-dependent and overspray losses are fairly high. Sprayed resist is made up of tiny droplets which strike the surface and flow outward to form a wet film. When too much solvent is present, excessive flow takes place and edge buildup is likely. With too little solvent present, the droplets do not flow and level adequately, and "orange peel" or uneven coatings occur. Ideally, immediately after application, there should be a smooth wet film, which will level with minimal flow during drying.

Spraying, like roller coating, covers only one side at a time. Immediately after it is sprayed, the part should be dried in a horizontal position for at least 5 min before coating the other side. Care must be exercised to avoid excessive overspray, which could damage the dried coating on the back side. Spray-coating techniques vary with operators and types of resist. Resists of low viscosity, such as AZ-340 or KAR 3 (11 to 12 cP), can be used as received at about 15 to 20 psig air pressure with good results. AZ-345 and Dynachem PRS-175 at 17 to 19 cP can also be sprayed as received or diluted four parts resist to one part thinner if desired. High-viscosity resists, such as AZ-111, may require dilutions such as one part thinner and three parts resist, although Shipley has indicated that AZ-111 could be sprayed as received (about 36 cP) using 30 psig spray pressure. When one sprays, clean filtered air or nitrogen free of oil and particulate matter must be used. Parts should be flat or within 30° of horizontal to reduce uneven coatings from flowing. After they are sprayed, parts should be allowed to dry in a horizontal position. Thinning, when necessary, should utilize the correct solvent for the best results. Shipley, Dynachem, and Kodak all have available thinners for their resists.

Equipment requirements for hand spraying are fairly simple. In addition to the propellant gas, a paint spray booth and gun are necessary, both of which are readily available commercially. Fixtures to mount, hold, and dry the parts within 30° or less of horizontal are required. The spray gun must be kept clean and any resist removed from it filtered prior to subsequent use. Dried resist particles may be codeposited with wet films or spray pattern. Distortion from clogging can occur without frequent cleaning of the spray gun. This is particularly true if different types of resist are used in the same gun and different solvent systems are involved. Application techniques may vary with operators while a satisfactory coating with regard to thickness and uniformity is still obtained. The actual technique is really unimportant as long as the results are satisfactory. A generalized method simply requires a series of horizontal passes of equal distance apart followed by a series of similar vertical passes. The idea is simply to ensure no missed areas without loss of uniformity. Each pass with the spray should produce a pattern that slightly overlaps the previous pattern to ensure good coverage. For the parameters mentioned herein, 10 to 20 cP resist viscosity and a pneumatic pressure of 15 to 20 psig, a good distance from the part is about 1 ft with a spray pattern spacing of 2 to 3 in. for each pass.

Good uniformity for hand spraying is 50 to 70 percent, with typical overspray losses of 50 percent. Better results are claimed for the Zicon spray unit; it utilizes the heated vapor from propellants which are liquid at room temperature.

Other Application Methods

Flow coating, the simplest method of application, and whirling are infrequently used. The former method consists of flooding about 60 to 80 percent of the area of the part with resist and tilting the part at various angles until all portions are wetted with the resist. Thereafter, the part is returned to the vertical and allowed to drain and dry. Problems arise from edge effects, where unusually thick coatings may form, and plated through holes, where flow through occurs. Chemical milling operations may employ large flow coaters wherein the photoresist is pumped through orifices over both sides of the vertically held part. Excess resist drains from the part back in the sump, where it is filtered and used again. Some flow coaters of this type are designed for continuous processing or conveyers and are equipped with dryers and prebake ovens.

Large whirlers used in the engraving and lithographic processes are equally adaptable for processing electronic parts without open holes. Compared with roller coating, spraying, or dip coating, whirling has a rather low output per unit time and a loss of resist comparable to or greater than the latter methods. During the application open holes pass resist which streaks outward on the bottom side when whirling starts; this limits the use of whirlers to flat parts without open holes or to parts on which a resist image is required only on one side. Still, any resist can be applied by this method and outstanding uniformity obtained. A coating factor of 90 percent should be readily attained. To use the whirler, the cleaned part is affixed to the spindle and the resist applied by one of several methods. Flooding the part prior to whirling is one method that gives very uniform coatings, but resist losses are high (85 percent or more). The method of direct resist application need not be uniform itself as long as it is rapid enough to leave a wet flowable coating. Pouring and spreading with a brush can be used. The important consideration is that for good uniformity whirling be started while the coating is still fully liquid.

After application of the wet film, whirling is started. Most macrowhirlers have low spinning speeds compared with microwhirlers. The large whirlers spin up to 150 to 200 rpm, whereas 8,000 rpm is common for microwhirlers. For resist viscosities of 10 to 20 cP, 80 to 100 rpm will give adequate results. Some whirlers have built-in heating elements that dry the part while spinning. A prebake should still be required; however, air drying can be eliminated.

AIR DRYING

Because of lower molecular weight, the positive resists have less tendency to retain solvents in amounts that could result in processing problems than negative resists. However, the coating is usually thicker than negative resists for the corresponding chemical resistance. The formation of a solvent-retaining skin on the resist surface is therefore possible, making air drying a requirement unless infrared prebaking is used. For most coating thicknesses, a 5- to 10-min air dry is adequate. Very heavy coatings may require 20 min or more. With infrared drying, only a short time (a minute or less) is required to allow the coating to level itself.

PREBAKING

Baking resist films prior to exposure, conveniently termed "prebaking," is essential in positive-resist processing. Prebaking removes the solvents, which could interfere with exposure, from the film. For reasons not readily apparent, the prebake parameters affect the pinholing tendencies of the image areas, also. Too low a prebake temperature (<60°C) may allow the unexposed resist image to be damaged by the developing solution and would tend to prolong developing times because of likely underexposure. Resist adhesion is also improved, and the film is hardened for greater resistance to handling damage. Excessively high prebake temperatures tend to degrade the resist film. In no instance should the prebake temperature exceed 110°C (230°F), preferably 100°C (212°F) for extended times. Some resists have greater thermal resistance than others, but 80°C (176°F) appears suitable for most of the commonly used positive photoresists. Times may vary. Heavy parts reach temperature more slowly than thin or small parts. Frequently opening the oven door may tend to keep the temperature low. Generally, 20 to 30 min at 80°C will give adequate results for these types of resists unless unusually thick coatings are used.

Infrared radiation is often used to prebake positive resists. The radiation penetrates the resist and heats the metal surface, which drives off the solvents from the metal base outward. There is less chance of a surface skim forming or solvents being retained with a heated surface beneath the resist. Of primary importance in the use of infrared ovens is to avoid heating the resist to an excessively high temperature. Variation of the power to the heating elements or the dwell time in the heated zone are methods of control. Conveyerized ovens vary dwell time by conveyer speed, and many do not have heating-element control. The temperature that is reached is a function of the dwell time in the heated zone at a constant heater output.

With infrared baking, times are much less than with other methods. For example, a swell time of 4 min may allow a peak temperature of 100°C to be reached, which would be adequate for moderate thicknesses of resists. Fairly high temperatures can also be tolerated for short periods of time with infrared heating. Moderate thicknesses of AZ-111 can be successfully prebaked at a 120°C (248°F) peak temperature with a total dwell of 3 to 5 min. Figure 5-4 shows a time-temperature cycle determined from

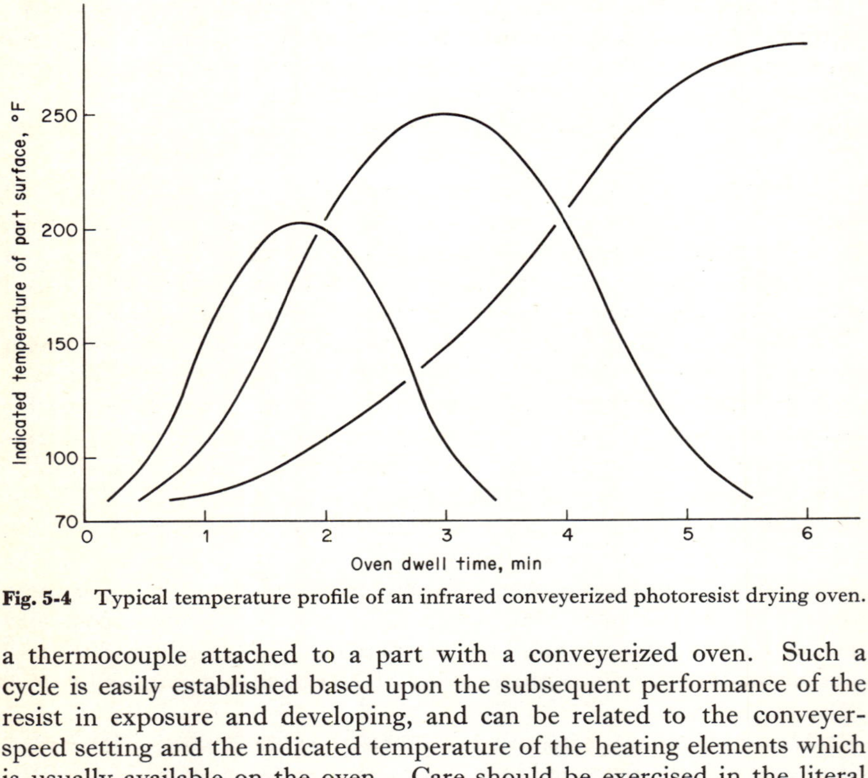

Fig. 5-4 Typical temperature profile of an infrared conveyerized photoresist drying oven.

a thermocouple attached to a part with a conveyerized oven. Such a cycle is easily established based upon the subsequent performance of the resist in exposure and developing, and can be related to the conveyer-speed setting and the indicated temperature of the heating elements which is usually available on the oven. Care should be exercised in the literal interpretation of photoresist manufacturers' data regarding heating-element temperatures unless the oven is the specific kind indicated by the data sheet. The use of a matrix may be helpful (Chap. 6).

Variations in temperature from part to part will occur with differences in surface finishes. Bright, glossy low-emittance metal surfaces will get hotter than parts with rough or nonreflective surfaces. Temperature differences of 20°F or more can occur simply because of surface finishes between parts. Similarly, variations in surface finish on the same part can cause hot spots where the resist may have become overheated. Hot spots can also occur because of a similar phenomenon on the radiating surface. Short of changing heating elements, there is little remedy. The effect can be dampened somewhat by increasing the distance between the part and heating element.

EXPOSURE

As with negative resists, exposure is likely the most important single step in the process. Unlike negative resists, however, exposure is not nearly as critical. Moreover, resist-image adhesion is not dependent upon adequate exposure. Positive-resist exposure establishes the image through a chemical reaction described in Chap. 2. The Sus reaction is shown in Fig. 5-5. Exposure changes the affected area from a resinous surface

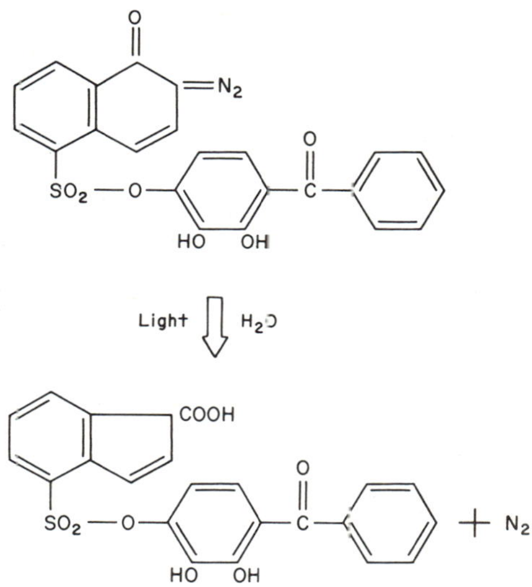

Fig. 5-5 Reaction of a positive-resist sensitizer with light. (Sus reaction. See Fig. 2-18*a*.)

insoluble in acids and dilute alkalies to a readily soluble chemical compound, thereby creating an image.

Underexposure simply will not produce an image in extreme cases, but it promotes scum formation in more subtle instances. Developing will be incomplete. Prolonged exposure tends to reduce line widths, but with less dependence upon coating thickness than negative resists according to the work of Tulumello and Harding (2). This is probably due to the effect of the developer on the negative resist and the nonpropagating nature of the exposure reactions of positive resists. The direct, visible results of overexposure will be a reduction in line width dependent upon the degree of collimation of the light source and probably the light intensity to some extent. However, the heat from the light may tend to cause scumming with some resists in the developed areas. Indirectly, scumming tends to increase the time in the developing solutions, which can promote pinholes in the image areas or, in extreme instances, physically damage the image. Because adhesion and exposure are not directly related in positive-resist processing, adhesion is very uniform on a clean surface.

Positive resists have photosensitive elements such as the ones shown in Fig. 5-6. It has been suggested that 1,2-naphthoquinone diazide is much

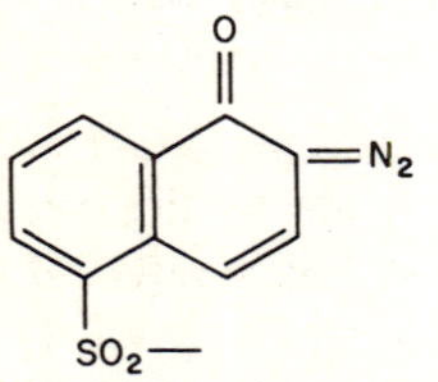

(a) Naphthoquinone(1,2)diazide(5)-sulfuryl radical

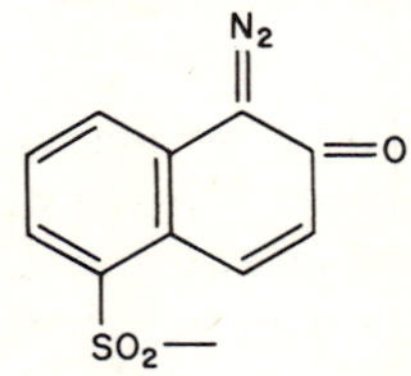

(b) Naphthoquinone(2,1)diazide(5)-sulfuryl radical

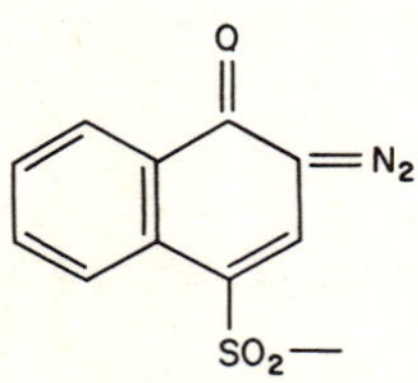

(c) Naphthoquinone(1,2)diazide(4)-sulfuryl radical

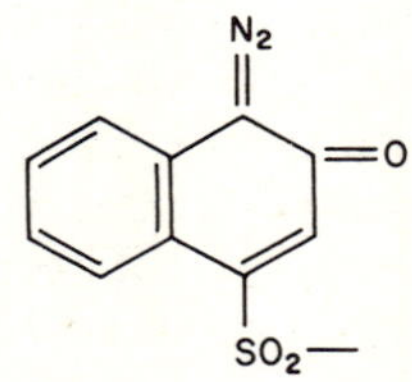

(d) Naphthoquinone(2,1)diazide(4)-sulfuryl radical

Fig. 5-6 Forms of naphthoquinone diazide. [Form (*a*) is believed to be the most photosensitive.]

more photosensitive than 2,1-naphthaquinone diazide and that the location of the sulfuryl ($-SO_2-$) linkage at the fourth or fifth carbon in the naphtha ring is also a factor.

From a practical standpoint, control of exposure can be exercised without knowledge of the actual aspects of the sensitometry. The film of resist is made up of a specific quantity of photosensitive elements, and a related quantity of light energy must be absorbed and reacted for solubilization to occur. Only absorbed radiation can react, and not all reactions are fruitful toward solubilization in the developer. Diazides absorb mostly in the narrow band between 3000 and 4000 Å. Peak sensitivity occurs around

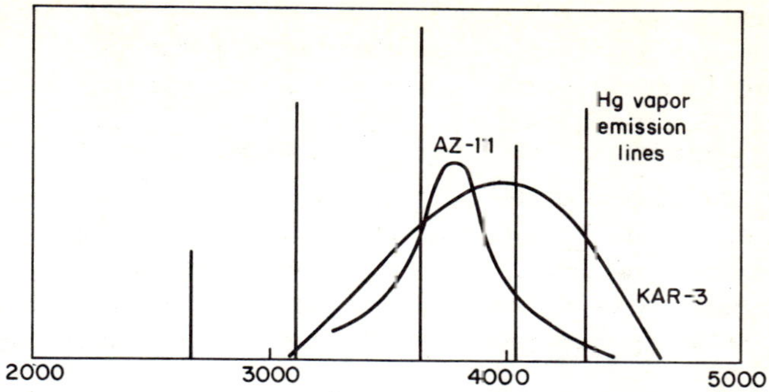

Fig. 5-7 Sensitivity of some photoresists in relationship to mercury-vapor emission lines.

3400 to 3700 Å for many compounds. Figure 5-7 gives a radiation spectrum of interest in the exposure of photoresists of this type relative to mercury-vapor emission lines. Safelights in the yellow, orange, or red range are adequate. Some resists can be used for plating and etching applications without the need for safelights in those process areas. In fact, resists such as Shipley AZ-119 or AZ-111 can be used for some acid etching even with the image fully exposed. In addition to mercury-vapor light sources, there are several types of light sources that can be satisfactorily used to expose positive resists. These include xenon, carbon-arc, and certain types of fluorescent lights. Chapter 4 contains a discussion of the various light sources and types of printers.

Basic methods of determining exposure times are similar to those used for negative resists. These include (1) step tablet, (2) calculation, (3) image accuracy, and (4) inspection.

Step Tablets

Step tablets (or sensitivity guides) are quick and sufficiently accurate to give acceptable exposure times both as a tool for initial evaluations and as a day-to-day control method. The only function a step tablet performs is to indicate visually the quantity of light that reacts with the photoresist (see Figs. 5-8 and 5-9). By varying the exposure time, different final or threshold steps will be obtained. As the positive resist is decomposed by light, the resulting step-tablet image is opposite from that obtained with the negative resists. Areas around the step tablet will be removed, and it is desirable to expose a portion of a photomask containing typical circuitry lines for inspection purposes along with the step tablet. The circuit image should be glossy and free of scum if the exposure time is adequate. Exposure times that will render the first six to eight steps soluble during developing are generally adequate for resists of this type.

(a) (b)

Fig. 5-8 Step tablets. (*a*) Kodak No. 2 step tablet. (*Eastman Kodak Co.*); (*b*) Stouffer 21-Step Sensitivity Guide. (*Stouffer Graphic Arts Equipment Co., South Bend, Ind.*)

The first step in determining the exposure time is the preparation of test panels coated with repeatable thickness of resist. All processing before and after exposure must be held constant during this or any of the exposure tests. The step tablet is placed emulsion side against the resist, and an initial exposure is made. Exposure time is not critical but should be short enough to ensure underexposure (steps 1 to 4). Subsequent exposure times are uniformly increased to a point of overexposure (step 8 or greater) to bracket the optimum time (step 6 or 7). For fixed-light-source printers with 1,000-W mercury-vapor lamps at a distance of about 3 ft, an initial

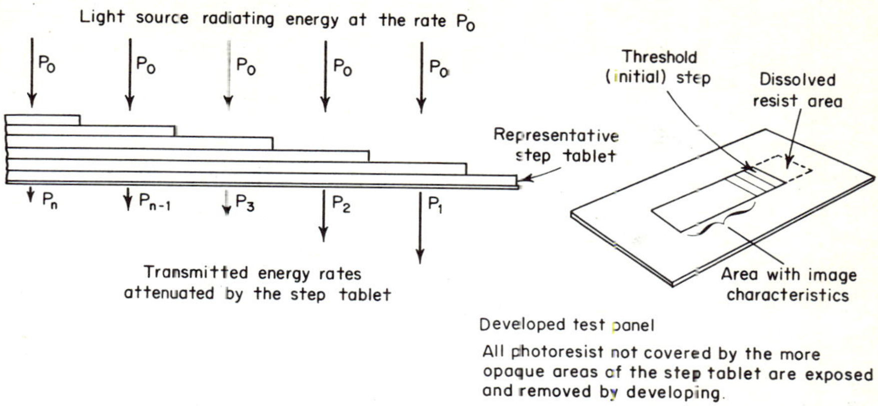

In this sketch, the changing transmission characteristics of the steps in the step tablet are represented by an increasing thickness of homogeneous layers. Radiant energy is delivered at a constant rate P from the light source uniformly over the surface of the step tablet. Each increasing step has an attenuating effects on P; it reduces the rate at which energy is transmitted. The quantity of energy Q that the photoresist beneath a given step will see is a direct function of the rate P_n at which the energy is being transmitted over the surface area of the step multiplied by the time t of exposure. The varying opacity of the steps reduces the intensity transmitted accordingly. As the time is constant for all steps, the energy that the photoresist receives will decrease with increasing opacity. A quantity of energy Q_m must be reacted by a given thickness of resist for solubilization to occur. When the value of Q_m is approximately equal to the energy received by the resist beneath the sixth, seventh, or eighth step, the exposure time is adequate for many types of positive resists. The condition of the resist at these steps does not indicate the condition of the image but rather serves as a visual indication of the quantity of light that the resist coating will see at that exposure time with that particular printer. Resist beneath the higher steps (12 or higher) is a good indication of the probable image condition.

Fig. 5-9 Step-tablet use with positive resists.

exposure time of 1 min would be a good starting point. Thereafter, simply doubling the time until overexposure is obtained is all that is required. Scanning and moving-light-source printers have a dial that can be divided into four or five increments. Exposure testing can be started at the fastest setting and terminated at the slowest setting, or a setting near the stalling point of the drive mechanism.

It is possible to use a single large panel in the tests and to keep the exposed and underexposed areas covered with an opaque material. A single developing step can thereby be utilized. A preferred alternative would be to prepare several panels and expose only once on each. This approach eliminates the danger of error from heat fogging. Both sides of a printing frame may be tested simultaneously by using two step tablets. If thin test specimens are used, the step tablets can be taped together (using clear plastic tape) at the transparent end with the emulsion sides facing. Exposure tests can then be made simply by straddling the part with the two step tablets and printing.

Calculation Methods

The proper exposure time can be estimated with reasonable accuracy by calculating from a known time and the resulting step. For example, if a step of 4 is obtained after an exposure time of 3 min, it would be possible to predict an exposure time to obtain a step of 6. This is possible because the light transmission changes uniformly and predictably with changing steps in a step tablet. For a 21-step tablet with 0.15 incremental differences, the corrected time t_c is related to the reference time t_R and the step obtained at the reference time S_R as follows, where S_c is the correct desired step:

$$t_c = t_R(\sqrt{2})^{S_c - S_R}$$

$$\begin{aligned} t_c &= 3(\sqrt{2})^{6-4} \\ &= 3 \times 2 \\ &= 6 \text{ min} \end{aligned}$$

That is to say that an exposure time of 6 min is required to obtain a step of 6 if 3 min gives a step of 4. This method is useful when determining exposure times on new or unfamiliar printers or with different photoresists. The proper exposure can quickly be zeroed in almost without regard to the error in the initial estimate. On the second exposure, it is generally necessary to run a piece of circuitry pattern along with the step tablet. The photoimage should be inspected for scumming, edge acuity, and general condition before accepting the exposure time as adequate.

Methods Involving Image Accuracy

Actual line-width measurements can be used to determine exposure times. The resist image is simply compared with the photomask at specific points. Measurements thus obtained are plotted and the most accurate time (where the deviation is 0) selected. However, the condition of the image must also be noted. An accurate image with scumming is of no value, obviously. An example using AZ-111 is shown in Table 5-4 and plotted in Fig. 5-10. By these criteria an exposure time of 6 min would be adequate.

Although the plot is for an etching pattern, it will probably be adequate for both plating and etching patterns. A plating-pattern plot from a similar photoresist would be the opposite and start with positive, or plus, deviation values. The point of intersection of the two curves would be near 6 min, but the intersection would be the most accurate single exposure time, regardless.

Resolution patterns (Fig. 5-11) have been used as a means of determining exposure times. Much as step-tablet methods, resolution-pattern methods require that a number of exposures be made until the desired results are

TABLE 5-4 Results of Line-Width Measurements of Positive-Resist Images

Point	Line-width deviation at exposure time (mil)			
Point	2 min	4 min	8 min	12 min
1	+0.8	+0.5	−0.4	−0.6
2	+0.7	+0.4	−0.6	−0.8
3	+0.8	+0.5	−0.5	−0.7
4	+0.8	+0.5	−0.5	−0.8
5	+0.7	+0.5	−0.5	−0.8
6	+0.7	+0.4	−0.5	−0.7
Average	+0.75	+0.47	−0.50	−0.73

RESIST: AZ-111. PHOTOIMAGE LINE WIDTH: 11.2 to 11.3 mils on 4-mil Mylar base. PRINTER: Millington with 1,000-W mercury-vapor source.

obtained as indicated visually by the pattern. A series of exposures is made with the resolution pattern at varying times. The condition of developed image and the finest space which is opened are the criteria evaluated. Overexposure will tend to open up fine spaces, but the fine lines will disappear. Conversely, underexposure will tend to have the opposite effect. The best time is the one that gives the finest combination of lines and spaces with an acceptable image condition.

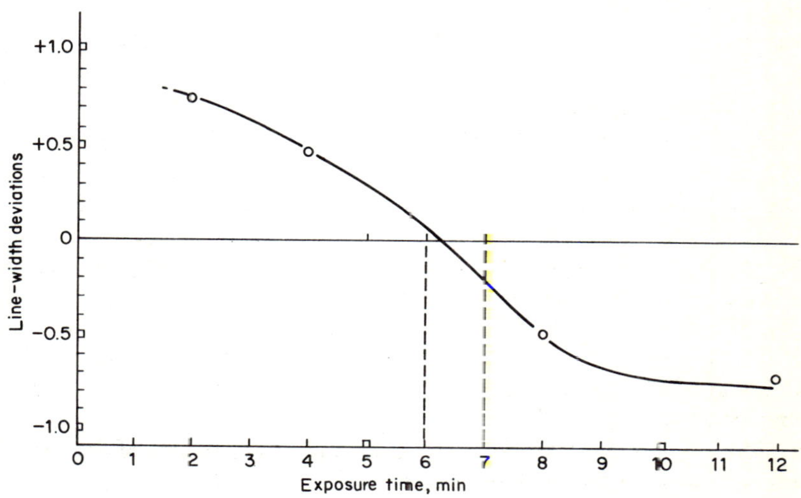

Fig. 5-10 Plot of line-width variations from Table 5-4.

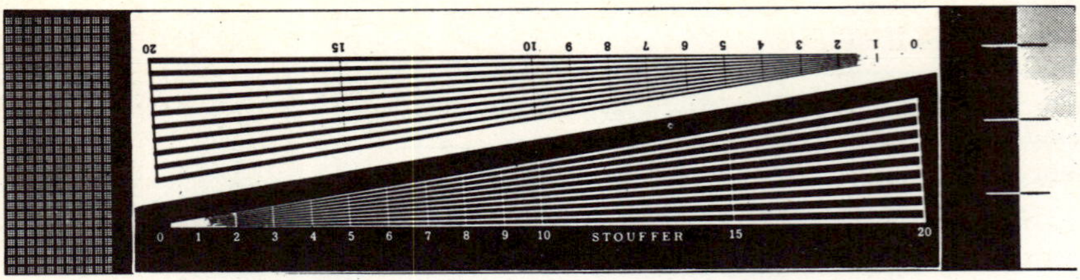

Fig. 5-11 The Stouffer Resolution Guide. (*Stouffer Graphic Arts Equipment Co., South Bend, Ind.*)

A method combining the step-tablet and line-width-measurement methods was proposed by Stienhoff (4). A line pattern of known width is placed beneath a step tablet and exposed for an extended period. After developing, the resulting line widths are then measured at each step where a pattern exists. The exposure time is calculated by determining the equivalent time that would be required to give the quantity of light that the step had received to the whole pattern.

Inspection Method

As the name "inspection" implies, exposure times can be estimated by those skilled in the art from the appearance of the resist image. Exposure times are varied using a number of test specimens (four to six) and a photomask. Time increments are varied from under- to overexposure or until a satisfactory-appearing image is obtained on several increasing exposures. Table 5-5 shows the result of an exposure test by inspection.

TABLE 5-5 Determination of Exposure Time by Inspection of Image

Time (min)	Result of inspection
2.	Incomplete image
4.	Image formed but resist still not completely developed
6.	Image developed cleanly with no visible damage
8.	Same as 6 min
12.	Same as 6 min

RESIST: AZ-111. PRINTER: Millington with 1,000-W mercury-vapor lamps at 3 ft.

A time of 6 to 8 min would be adequate from the data.

DEVELOPMENT

Areas exposed to light for a sufficient period of time become soluble in the developer. The change, described previously, results in the formation of carboxylic acid groups in the sensitizer molecule, which form alkaline-metal or amine salts in the developing solution and are thereby dissolved. Unexposed areas have little or no affinity for the developing solvent, which is in contrast to the negative resists, where both the exposed and unexposed areas tend to attract and retain solvents. Retained solvents tend to cause swelling and pattern distortion, which does not occur to any extent with positive resists. It is possible that a reaction does occur in the unexposed areas, however. Since most positive-resist formulations contain phenol resins and diazide sensitizers, it is possible that a coupling reaction takes place.

Coupling has long been known in the formation of azo dyes. Coupling occurs at a very rapid rate at the proper pH with the formation of an azo linkage between the sensitizer and phenol resin. At a neutral or acid pH, quinone diazides are stable and resistant to coupling. Excessive alkalinity retards coupling through the formation of other compounds. At the proper pH range, however, coupling could serve to reduce the loss of the phenol resin from the surface.

Developing solutions contain, at the least, a source of alkalinity and a surfactant in an aqueous medium. Additionally, buffering agents, glycol ethers (5), and coupling agents may also be present. As the solution is used, the carboxylic acid groups formed by the exposed sensitizer will be neutralized by the alkali substance present in the developer.

Salts thus formed are dissolved in the developing solution. As a result, the pH of the solution will drop with use. Stockley (6) indicated that the normality of Shipley AZ-303 could be used as a control indicator. If one starts at 0.33 *N*, the recommended minimum is 0.29 *N* before replacement. The Shipley AZ-303 developer is supplied as a concentrate and diluted with water. It is used with most of the resist except for AZ-1350 and AZ-1350J. Carbon dioxide absorption from the air also consumes free alkalinity. It is not generally a severe problem in strongly buffered developers, such as AZ-303, but does affect dilute solutions, such as AZ developer.

Temperature has some effect on development. Generally, temperatures in the low or mid-60s or above 80°F may tend to cause changes in developing times. Higher temperatures may cause problems with resist damage, whereas incomplete developing or scumming may occur at low temperatures. Spray developing gives better results at lower temperatures than immersion developing, but both operate well between 68 and 80°F.

Spray developing is an excellent way to develop positive resists. Spray developing gives superior results to immersion developing with either positive or negative resists. Cleaner, more thorough, more rapid removal of the exposed areas (or unexposed for negative resists) is obtained with less damage to the image by proper spray developing. Because of the somewhat hazardous nature of negative-resist developing solutions, it is not widely employed there, but the newer dry-film (Chap. 6) negative resists utilize it exclusively with special equipment. As the developer for positive resists in aqueous base, there is a wide variety of compatible materials available. Further, the extensive ventilation systems necessary for solvent-base developers are avoided because of the developers' low volatility. Used etchers of polyvinyl chloride (PVC) can be used, and special equipment is also available. Both normally contain spray-rinse chambers after developing and can be equipped with either an air knife for termination or attached directly to etchers for further processing.

One problem that occurs with spray developing is foaming. The surfactants employed are very active, and the agitation afforded by spraying causes extensive foaming. Various foam suppressors are available, such as aliphatic mineral oils, but some care should be exercised in the selection. The material should not cause subsequent processing problems in plating or etching, nor should it damage the resist image. A specific foam-suppressing formulation is available from the Shipley Company.

RINSING

The primary purpose of rinsing when processing positive resists is to remove water-soluble substances from the part. Unlike negative resists, which require a forceful rinse to remove the insoluble solvent-logged portions of the image, the reaction products of the exposed areas of positive resists are water-soluble acid salts. Immersion rinses can readily be used, although spraying is preferable. Water purity does not have a significant effect on the rinsing operation, although the possibility of calcium in the rinse water forming acid salts of low solubility has not been investigated. Scumming would be the likely result of such a reaction if it occurred.

EXAMINATION AND TOUCH-UP

While it is possible that very careful inspection of the photomask and a high level of cleanliness during the processing could eliminate the need for examination and touch-up, most shops find these operations necessary. Depending upon the cost of the part and its allowable design tolerance for defects, it may be necessary only to examine parts and set aside or reprocess those with defects. Examination of the photoresist prior to postbaking is an effective way to maintain some degree of process control. By locating reoccurring defects for correction before large numbers of defective parts are processed, overall process control is exercised on a "cure-and-ailment" basis. Proper process control, however, can prevent such defects from occurring.

Touch-up consists of fortifying the resist image in defective areas and removing extraneous resist that would interfere with subsequent processing from the nonimage areas. Positive-resist images are generally less resistant to solvents than those of highly polymerized negative resists, and the same type of paint or touch-up solution may not be useful in both instances. In fact, touch-up resins in active solvents, such as acetone or MEK (methyl ethyl ketone), can actually damage the image by dissolving the resist in the touched up area. A "fisheye" type of defect can result, with the resist becoming very thin at the perimeter of the area. Resins dissolved in solvents of very low polarity are generally usable in this respect. In addition to having a compatible solvent system, the touch-up ink must adhere well to the resist and to metal and be of sufficient contrast in both instances for good visibility. Finally, the ink must itself have adequate chemical resistance. A desirable feature is long pot life. Inks that dry quickly are often wasted because the scanning time to locate defects allows them to harden or skim in open containers. Asphaltum, some screening inks, and Shipley touch-up material have been used for resists in this category.

To remove extraneous materials from the metal portion of the part, instruments such as X-acto knives are often used. Some prefer non-metallic instruments, such as the orange sticks used in fingernail care. Regardless of the instrument, locating thin films of scum or extraneous resist is usually more of a problem than the removal operation. When parts are to be etched, a quick pre-touch-up pass through the etcher may offer greater contrast in the metallic areas.

POSTBAKING

Postbaking improves the adhesion and chemical resistance of the image. For most applications, postbaking is desirable for positive resists and is usually essential when the resist is used for pattern plating. The temperature resistance of positive resists is less than negative resists and postbake temperatures correspondingly lower. Excessively high temperatures tend to initially cause resist flow, then embrittlement, then thermal degradation. Flowing is usually not serious enough to cause problems for macropatterns. Embrittlement can be a problem, however, particularly when the parts can be flexed, such as multilayer details. Adhesion loss can also accompany embrittlement in some instances. Generally, resists such as AZ-119 or AZ-111 can be safely postbaked at 80 to 90°C with good results. Times for etching are as low as 10 min with longer times, for example, 20 or 30 min, for plating. Dynachem does not require a postbake by the literature, but a short bake cycle similar to the Shipley resists may be useful for plating. Shipley AZ-340 and AZ-345 are apparently more temperature-sensitive than other AZ resists. A postbake temperature of 60 to 70°C can be used for up to 30 min. KAR-3 can be postbaked similarly to Shipley AZ-119 or AZ-111 resists. All resists can withstand higher temperatures, but stripping will be more difficult.

STRIPPING

Positive resists strip comparatively easily in alkaline solutions after the image is reexposed, and in a variety of organic solvents, such as ketones, amines, or glycol ethers (Table 5-6). Ester solvents, such as butyl

TABLE 5-6 Some Stripping Solvents for Positive Resists

Amines	Glycol ethers	Ketones	Others
Dimethyl formamide	Cellosolve	MEK	Dioxane
N-Methyl-2-pyrrolidone	Methyl Cellosolve	Acetone	Sodium Phenolate and cresols
Diethanolamine	Butyl Cellosolve	Methyl isobutyl ketone	Shipley 1112 stripper
Triethanolamine	Butyl Carbitol	Cyclohexanone	
	Cellosolve acetate		
	Methyl Cellosolve acetate		
	Butyl Cellosolve acetate		

acetate, can be used on the resist with a light brushing, which is useful in all instances. Some of the solvents, such as DMF (dimethyl formamide), may be harmful to epoxies and should be used on plated parts with no exposed epoxy areas. Mixtures such as butyl Cellosolve and *N*-methyl-2-pyrrolidone (GAF) make useful soak strippers that are water-soluble and of low volatility. Perhaps the most widely used strippers are MEK or acetone (dimethyl ketone). In spite of the flammability and volatility, their stripping action is excellent. MIBK (methyl isobutyl ketone) is less volatile, but odorous. Cyclohexanone is a good solvent with low volatility. All the Cellosolve series are good strippers. Cellosolve acetate is used in some of the resist formulations as the true solvent. Dioxane is a good resin solvent, but only inhibited grades should be used.

Concentrated mixtures of caustics and phenol derivatives in water will rapidly strip most resists in this series when heated. Besides being odorous, however, phenols are very difficult to completely remove from circuit-board resins. Shipley 1112 stripper is excellent.

Commercial methylene chloride base strippers and alkaline strippers effectively remove most positive resists. Regardless of the solvents used, all are toxic to some degree and care must be exercised. Breathing solvent fumes or contact with skin and eyes is always harmful and often serious. Gloves should be tested for compatibility before use and proper ventilation provided. Solvents that are readily absorbed through the skin, such as dimethyl sulfoxide, should be avoided.

STORAGE AND HANDLING

Unlike the negative resists, positive resists do not lose their photosensitivity after processing. Some resists, such as AZ-111 or AZ-119, retain good chemical resistance to acids and mildly alkaline solutions even though the image is exposed to actinic light sources. Other may tend to degrade in ordinary white flourescent lights, although most are good for several hours in normal shop lighting. Tests to determine the effect of various storage times under shop-light conditions are advisable if an unacceptable incidence of image defects is experienced. Pinholes may be the resulting defect in some instances.

Resists in the liquid form tend to slowly decompose at room temperature. Apparently, the sensitizer evolves nitrogen and loses photosensitivity. Bottles that have been stored unopened for several years at room temperature may actually spray resist or noisily release gas under pressure when opened. Nitrogen bubbles may be present in the resist when it is applied. Although it is comparatively rare that the photosensitivity is lost altogether, erratic results during processing may be experienced. Often, the resist may be filtered and used. The nature of the reactions in the solution which causes the problem is not known. Quite possibly, the sensitizer undergoes some form of coupling or dimerization, which may cause precipitation as well as gassing.

As with any wet-film resist, contamination is a problem. Chapter 4 contains a discussion of contamination problems that are applicable to both types of wet-film resists. Table 5-7 shows some additional sources.

TABLE 5-7 Some Sources of Contamination

Operation	Typical contaminant
1. Resist application	Water, abrasives from cleaning, fingerprints from handling after cleaning, matter from unfiltered resist, fibers and particles of copper from board previously processed (roller and dip coating)
2. Drying and prebaking	Airborne particles, fingerprints, actinic light, moisture, smog
3. Exposure	Airborne particles, fingerprints, dried resist on vacuum frame and photomask
4. Developing	Excess resist in developer, salt precipitates, residual wetting agents
5. Rinsing	Insoluble particles, metal salts

REFERENCES

1. G. Kogel and H. Neuenhaus, U.S. Patent 1,444,469 (1923).
2. J. Tulumello and W. B. Harding, *Plating*, vol. 54, no. 11, pp. 1234–1238, 1967.
3. T. L. Stienhoff, C. Isaacson, and G. J. Roelants, U.S. Patent 3,535,157 (1970).
4. T. L. Stienhoff, *PC-E*, November, 1966.
5. F. Enderman and W. Neugebauer, U.S. Patent 3,148,983 (1964).
6. F. Stockley, *Proc. EIPC Conf. Printed Circuit Manufacturing*, Hargen, Switzerland, February, 1970, pp. 93–107.

CHAPTER SIX

Dry-Film Photoresists

Although relatively new (about 1968) to the electronics industry, photosensitive dry films (1,2) have been used in the printing industry since the early 1950s. This is in keeping with the development of other photoresists which had their origin in the printing industry. Unlike the wet-film resists, however, the transition from the printing to the electronics industry was not as clear as the wet films. It was natural to think of wet-film photoengraving resist as an etch resist for printed circuit boards. The dry films were prepared and originally supplied on a rigid backing for relief printing as they are today. It took a thorough assessment of the shortcomings of the wet films and very original concepts to conceive the fundamental idea of dry-film resists. Since their introduction by Du Pont through the patent of Celeste (3), a large portion of the photoresist users have switched to dry films; an extraordinary growth and acceptance in the very competitive field of photoresists. Shortly after Du Pont marketed the Riston dry films, Dynachem Corporation introduced the Laminar series of dry-film photoresists. In 1970, W. H. Brady Company introduced the Phodar series but shortly removed it from the market.

The dry-film photoresist composite used by all dry films is shown in

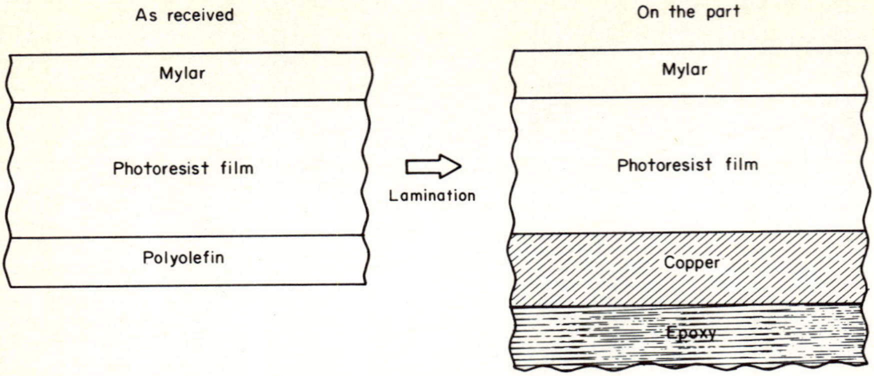

Fig. 6-1 Dry-film photoresist composition.

Fig. 6-1. From the start, dry films offered many advantages over the wet-film resists. Processing was drastically reduced both stepwise and timewise. Coating irregularities associated with wet-film resists were virtually eliminated.

This chapter will discuss the types of resists followed by processing information for the resists discussed. In addition to basic processing information, discussions on baking, touch-up, stripping, multiple coatings, and liquefied forms of the dry films will be included.

TYPES OF RESISTS

There are three basic types of dry-film resists; they are best categorized by the manner in which they are developed. These are (1) solvent-developing resists (Type I), (2) aqueous-developing resists (Type II), and (3) peel-apart resists (Type III). Types I and II give negative images and are processed in a conventional sense. The third category is unique in its processing techniques and gives a positive image. No liquid developing is employed. Table 6-1 gives a list of some resists available from the two major manufacturers by category. The term "tenting grade" in the table refers to resists which can be used for "tenting," or sealing off holes during etching. Figure 6-2 shows a tented hole.

Solvent-developing Resists—Type I

Type I resists are the most widely used of the dry-film resists. These resists are all developed in methyl chloroform (1,1,1-trichloroethane) at a temperature of $65 \pm 5°F$. From the basic patent of Celeste (3), there are essentially two types of resists in this category. These are a combination of a photosensitive resin with unsaturated acrylic functional groups and,

TABLE 6-1 Types of Dry-Film Resists

Resist	Manufacturer	Nominal resist thickness (mils)	Use
		Type I. Solvent-developing	
Laminar HS	Dynachem	1.5 and 2.0	Tenting grade
		1.0 and 1.3	Plating and etching
Laminar Orange		0.5	Plating and etching
Riston 6	Du Pont	0.75	Plating and etching
Riston 15S		1.5	Tenting grade
Riston 16S		1.8	Tenting grade
Riston 30S		2.9	Tenting grade
Riston 14		1.35	Plating and etching
Riston 25		2.5	Tenting grade
		Type II. Aqueous-developing	
Laminar A	Dynachem	1.0 and 2.0	Tenting and general use
Laminar AP and AX	Dynachem	1.0 and 2.0	Tenting and general use
Riston 218	Du Pont	1.8	Tenting grade
Riston 211	Du Pont	1.0	Plating and etching
		Type III. Peel-apart	
Riston III	Du Pont	....	Etching

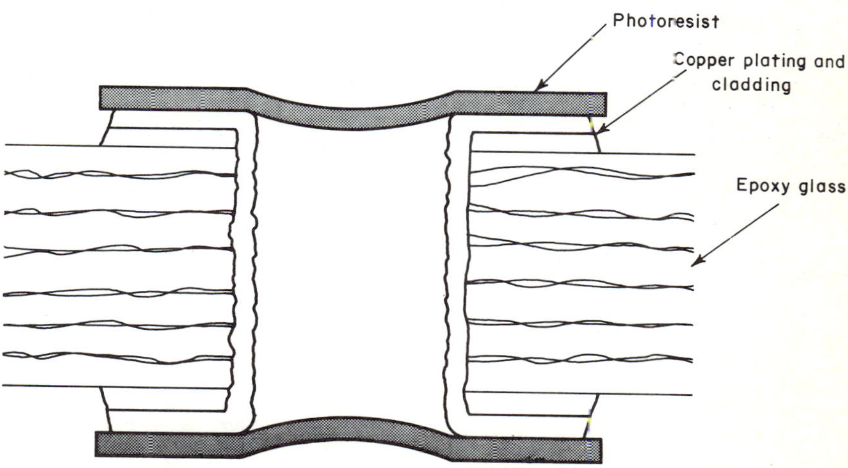

Fig. 6-2 Tenting a plated through hole with dry-film photoresist.

secondly, a combination of a polyfunctional unsaturated monomer and a nonphotosensitive binder resin. The components, from patents, are shown in Table 6-2.

TABLE 6-2 Basic Dry-Film Resist Components

Type A components	Type B components
Photosensitive resin	Nonphotosensitive binder resin
Inert filler (nonphotosensitive) resin	Polyfunctional ester monomer of acrylic acid
Sensitizer (photoinitiator)*	Polymerization inhibitor (for monomer)
Chain transfer agent	Sensitizer*
Thermal polymerization inhibitor	Dye*
Dye*	
Plasticizer	

* Sensitizer may contain a dye.

The first type of resist utilized a macro-polymer molecule which is photosensitized by virtue of acrylic ester linkages in which the acrylic unit is unsaturated. Figure 6-3 shows some of the examples from the patent of Schoenthaler (4), which was cited by Celeste.

The photosensitive Schoenthaler resin had variable compositions and was prepared from acrylate monomers, as previously indicated. For photosensitivity, a monoglycidyl acrylate monomer was included with other acrylate monomers. Monoglycidyl acrylate is formed by reacting

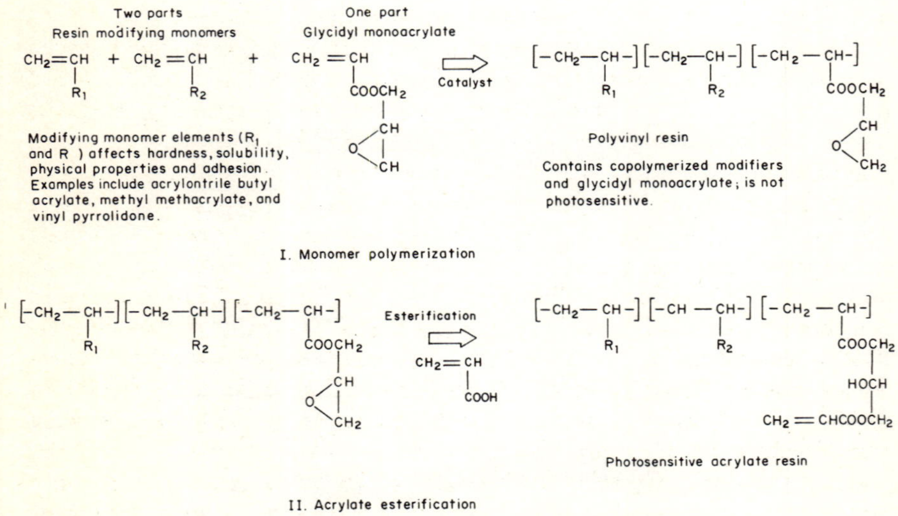

Fig. 6-3 Example of the photopolymer of Schoenthaler (4).

1 mol of acrylic acid with 1 mol of glycerine, which has three available hydroxyl groups for esterification. After esterification, two free hydroxyls are available for further reactions. The unsaturated group in the acrylic acid is reacted with other acrylic monomers to form a linear chain. Dangling from the chain are the esterifiable functional groups from the glycerine molecules. At this point, the resin is not photosensitive and contains no ethylenic unsaturation. Photosensitivity was gained by esterifying the glycerine functional groups with acrylic acid. Examples of monomers polymerized with the glyceride to form the linear chain included methyl methacrylate, acrylonitrile, vinyl pyrrolidone, and butyl methacrylate. At least 10 wt percent of the glycidyl groups is required to achieve adequate photosensitivity after esterification. If the level approached 100 percent, the polymer solubility characteristics, cost, and other properties reached undesirable levels. The glycidyl monomer was present in 25 to 33 percent levels in the examples cited.

In addition to the photosensitive resin, some formulations contained an inert filler resin which was not photosensitive. About two parts photosensitive resin to one part filler resin was the ratio suggested. The function of the filler resin was not made clear. A copolymer of methyl methacrylate and β-hydroxyethyl acrylate was given. Possibly, improved adhesion and reduced cost would be gained by the admixture.

The linear polymers are insolubilized during exposure by cross-linking the long chains into a very high-molecular-weight network through a reaction between the esterified acrylic functional groups, resembling in this respect the polyvinyl cinnamate resin used in the Kodak KPR resist series (see Chaps. 2 and 4) which was also patented by Celeste (34).

The second type of resist contained a mixture of polyfunctional acrylic or methacrylic monomers and a binder resin which was apparently not photosensitive. Dynachem (5) may use a mixture of unsaturated diallyl phthalate prepolymers as a binder resin admixed with acrylic monomers. Polyfunction monomers cited were pentaerythritol triacrylate, diethylene glycol diacrylate, and triethylene glycol diacrylate among others. These compounds are shown in Fig. 6-4. As may be surmised, these monomers are capable of forming highly interconnected three-dimensional network polymers upon exposure. The monomers were dispersed in the binder resin at a concentration as low as 21 percent to over 50 wt percent of the polymer. Binder resins cited in examples were generally mixtures of methyl methacrylate and methacrylic acid. Molecular weights cited were 30,000 to 60,000. In all, approximately 16 different types of binder resins were mentioned, covering most common resins. An exception was allyl phthalate resins. Phthalic acid esters of polymethylene glycol were mentioned, but no allyl resins. With this flexibility, a wide variety of physical properties could be obtained by mixing and varying binder resins.

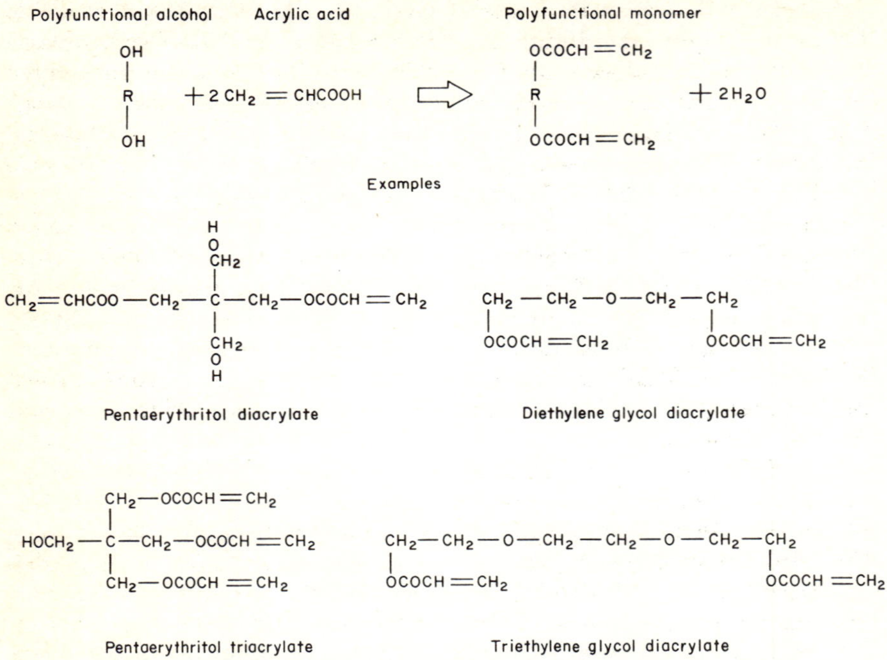

Fig. 6-4 Photosensitive polyfunctional monomers for use with binder resins.

However, in a later patent, Hurley and Lazarus (37) mentioned only polymethyl methacrylate, with and without free methacrylic acid in the polymer, and chlorinated rubber.

Both acrylic and methacrylic acid esters form photosensitive monomers. From the patent of Celeste and Bauer (6), pentaerythritol triacrylate esters are considerably more reactive and photographically faster than the methacrylate esters. Prior to exposure, monomers function as plasticizers to impart the desired flexibility and hardness to the film. The monomer physical properties as well as the ease of polymerization influence the ratio of monomer to binder. Celeste and Bauer (6) indicated that the glycol acrylate monomers tend to overplasticize. This causes soft films, which limits the amount of monomer and the ratio. Pentaerythritol triacrylate produces a much harder film under the same conditions and offers a large ratio. There were also other esters of methacrylic and acrylic acid. Dipentaerythritol and trimethylolpropane esterified with acrylic acid were also prepared and produced satisfactory results. Mixtures of monomers could be used to obtain the desired plasticizing action for the unexposed film. Trimethylolpropane triacrylate and triethylene glycol diacrylate were used in combination (37).

Two basic types of sensitizers were patented for use with the dry-film resists by Celeste (3). One type was a series of polynuclear quinones and the other was dye sensitizers. Upon absorption of actinic radiation, the quinone molecule undergoes complex reactions and forms short-lived, reactive compounds with an odd number of electrons called "free radicals." Free radicals generated by the sensitizer react with the unsaturated linkages in monomers and polymers. The overall result of a complex series of reactions is that one of the two pairs of electrons shared by two carbon atoms which make up an unsaturated double bond becomes reactive enough to form electron-sharing covalent bonds with other reacted double bonds. During these reactions, monomers join to each other, grow, and are thereby consumed. Further, monomer free radicals may also react directly with binder resins even though the latter may have no unsaturation. Electron pairs shared by reacted double bonds extended between linear polymers can cross-link them together chemically to form giant molecules. It is possible to have light of the correct frequency react directly with the double bonds to form free radicals, but the frequency range required is high and normally absorbed by glass, and a high-energy light source is required.

The polynuclear quinone sensitizers mentioned were compounds made up of several ring structures with two carbonyl structures in one ring. Many were in the patent of Notley (7). An example is 9,10-anthraquinone. In general, quinones react directly with light in the 3000 to 4000 Å range to form a pair of radicals. Chang (35) and Chambers (36) patented suitable sensitizers, also.

Dye sensitizers such as those that may be used in the resists (9–13) are a combination of dyes and reducing agents. The dye initially absorbs light energy and reaches an energy-rich, or excited, state in which no chemical change has actually taken place. In the excited state, the dye molecule is capable of initiating reactions that would not normally be possible. In this state, the dye reacts with the reducing agent to generate free radicals, which, in turn, react with the ethylenically unsaturated compounds. If monomers are present, chain growth starts and continues until termination occurs. Not all the reactions that occur promote monomer polymerization or cause polymer crosslinking. Such reactions include dissipation of the excited-state energy in the form of heat, fluorescence transfer to a nonreactive molecule, and others.

Free-radical polymerization can be terminated by oxygen. If present in sufficient amounts, it could interfere with adequate cross linking or prevent monomer polymerization to the point where developing could not be endured. Heiart patented the use of gas-impermeable films over the resist to preclude oxygen effects (14).

An example of dye sensitization is the combination of methylene blue and triethanol amine. Alles (15) used 0.2 and 1.5 g, respectively, for polymerizing approximately 10 g of pentaerythritol triacrylate. Methylene blue gave a visible image after exposure, but erythrosin did not, indicating that latent images did not necessarily form with all dyes. No latent image was mentioned with quinone sensitizers, although when used with the methylene blue dye and electron donor, the photographic speed was drastically increased and a latent image obtained. The assumption has been made in the processing notes herein that if a latent image was obtained after exposure, dye sensitizers were used, and a hold time after exposure was desirable.

Dyes are frequently sensitive to light in the visible range (4000 to 7000 Å), which would be undesirable in many instances. It is apparently possible to combine dyes and UV absorbers to filter out undesirable frequencies and create a discrete window. This would also reduce problems from reflected light.

Another ingredient was the chain transfer agent. More rapid exposure times and improved adhesion were reportedly obtained using this compound, but reasons were not given. No chemical unsaturation was present in the compounds cited, and they could be used with either of the basic types of photoresist (Table 6-1). The compound frequently cited was triethylene glycol diacetate because of its low vapor pressure, which retarded losses from vaporization. An ester of polyethylene glycol monosuccinate was also mentioned (15) for use with pentaerythritol triacrylate monomers, but the glycol diacetate was mentioned by Celeste for both basic compositions. About 10 to 15 wt percent was used based on the photosensitive ingredient. It also had a plasticizing effect.

Vinyl polymerization inhibitors were used during the esterification reactions in which the unsaturated acrylic and methacrylic acid were added for photosensitivity. These inhibitors may be retained to prevent cross linking and monomer polymerization reactions from taking place during heating. For pentaerythritol triacrylate, *p*-methoxyphenol (6) was used in concentrations of about 0.2 wt percent. Schoenthaler (4) indicated that small amounts of copper metal or salts as well as organic inhibitors prevented cross linking during esterification of the free hydroxyl groups on the glycerine (Fig. 6-3).

Celeste (3) did not mention adhesion promoters per se. Hurley and Lazarus (37), however, patented similar resist compositions containing photosensitive adhesion promoters for copper substrates. Heterocyclic nitrogen compounds mentioned included benzimidazole which was used in combination with tertiary 2-butyl anthraquinone. The compounds cited could react chemically with the copper surface. Roos (38) patented compounds that improved resist adhesion to silica, silicon, and ceramic surfaces.

Aqueous-developing Resists—Type II

Type II resists are characterized by their ability to be developed in aqueous (Dynachem Laminar A) or partially aqueous (Riston II and Laminar AX) solutions. There are several advantages in using these resists, especially for small shops, labs, and facilities that want to evaluate or use dry films but need to minimize initial installation expenditures. This is not to infer that these films cannot be used for high-volume applications just as well as the solvent-developing resists but rather to indicate the unique advantages for small operations. One major drawback is the limited resistance to alkaline solutions (pH >8.5), which must be taken into account in subsequent processes. However, with the proper control most resists can be used with pryophosphate plating solutions if the pH is controlled and the current density is low enough (for example, 20 A/ft^2).

The resins used in the resists contain acidic groups that react with free alkali ions to form a salt. This converts the resin into a water-soluble substance if the reacting groups are sufficient in number to overcome the effects of the hydrophobic elements in the polymer chain. As one may note, the developing reaction thereby involves an actual chemical reaction rather than the physical action of dissolving the resin in a solvent. The reaction may be represented by

$$\left(\begin{matrix}-R_1-\\ |\\ COOH\end{matrix}\right)_n(-R_2-)_x + n\text{NaOH} \longrightarrow \left(\begin{matrix}-R_1-\\ |\\ COONa\end{matrix}\right)_n(-R_2-)_x + nH_2O$$

where R_1 represents the repeating unit to which the acid group is attached and R_2 represents hydrophobic elements in the chain. For example, R_1 and the carboxylic acid group could be polymerized methacrylic acid and R_2 hydrophobic polymethyl methacrylate if the resin is not photosensitive. The ratio of these groups will determine to a great extent the ease of development, as well as the alkali resistance. Alles (15) used a 90 : 10 mol ratio for the resin just mentioned. Celeste (16), using a photosensitive resin, stated that a minimum of 5 mol percent was required for good developing in alkalies. From examples in the patent, the resist was actually developable in warm water when two of the three groups were carboxylic acid.

Monomers vary in hydrophobic tendencies and concentration. Either of these factors will affect developing time. Examples of monomers listed for aqueous developing films included pentaerythritol triacrylate (15–17) and polyglycol diacrylate compounds (1). Formulations citing a high concentration, for example, 40 wt percent, of pentaerythritol triacrylate generally contained some type of organic solvent in the developing solution, notably butyl Cellosolve (also called *n*-butoxyethanol or ethylene glycol monobutyl ether) in concentrations of about 6 percent, although there were claims of 1 to 50 percent. Among the monomers mentioned for use with completely aqueous caustic developing solutions were diethylene and triethylene glycol diacrylates.

Most other ingredients patented for solvent-developing resists were also mentioned. Either ketone or dye sensitizers could be used.

Peel-apart Resists—Type III

The operating principle of Type III resists resembles that of Du Pont Crolux films. When light reacts with the resist resin, it polymerizes, and the adhesion to the clear-plastic overlay is increased. The exposed resin undergoes a change in physical properties because of an increase in molecular weight. In the unexposed areas, the resin, of course, remains unchanged. After exposure, parts are developed thermally with the overlay in place. Initially, the part is heated. Heating is sufficient to cause flow in the unexposed areas but not in the polymerized exposed areas. While warm, the clear-plastic overlay is peeled from the part. By virtue of the adhesion to the Mylar material caused by polymerization, the exposed image is pulled from the metal surface by the departing plastic overlay. In the unexposed areas, the heat-softened unpolymerized resin undergoes a cohesive failure with the application of the peeling force. About half of the resin remains attached to the Mylar overlay and is peeled off with the exposed image. The lower half remains attached to the metal substrate and forms the image. Because the unexposed resin remains, the image is positive. The image profile is somewhat distorted from the tear-out of the exposed image, but it is usable for many applications. If unexposed, the image remains unpolymerized and readily removable. For greater durability, the image can be exposed.

There is little information regarding the composition of these resists. In an early patent, Burg (18) described thermoplastic resins admixed with polyethylene glycol diacrylate monomers, sensitizers, pigments, and antioxidants capable of thermal transfer. Transfer systems utilizing the pentaerythritol triacrylate monomers were mentioned by Celeste and Bauer (6).

Schoenthaler (4) in example XXV of his patent disclosure described a resin composition that could be thermally transferred to copper-clad circuit-board material after exposure and could resist etching in ferric chloride. The resin is shown in Fig. 6-5. From the example cited, the resist has the following weight percent composition:

Photopolymer	54.7
Filler resin	26.4
Triethylene glycol diacetate	11.9

Quinone sensitizer..................	6.5
Thermal polymerization inhibitor	0.5

The resin was dissolved in methyl ethyl ketone and cast on a Mylar film with an anchoring layer as given in the patent of Alles and Sauer (19). A second layer of Mylar was laminated in place at 127°C over the photosensitive resin. The side with the anchor layer was exposed and the image heated and transferred using the method of Heiart (14). Both circuit

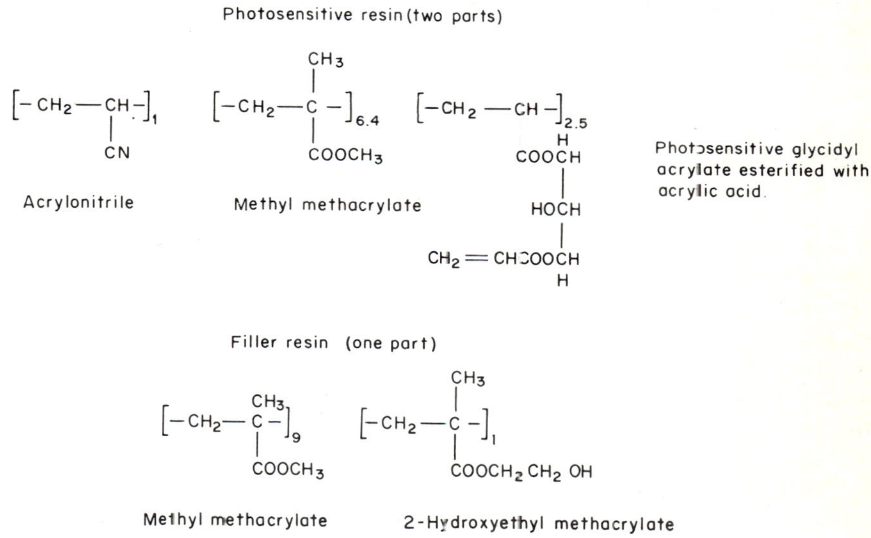

Fig. 6-5 Thermally developing dry-film resist-resin mixture.

boards and lithographic plates could be produced by this method. Du Pont (20) also patented a thermal-transfer image that was substantial enough to function as a screen-stencil image.

Processing information for these resists will not be discussed and is concluded at this point.

BASIC PROCESSING

The basic processing sequence used for both solvent- and aqueous-developing resists is similar. Both will be discussed in the remaining portion of the chapter. The process involves only cleaning the part, laminating the resist in place, exposing, and developing. Touch-up, stripping, and optional processes will also be discussed.

Part Preparation and Cleaning

In addition to the basic cleaning of the parts described in Chap. 3, there are three additional considerations when one uses dry films. These are:

1. Parts must be very clean.
2. Parts should be warm at the end of the cleaning operation.
3. Fine surface defects in the part surface must be avoided.

Wet-film resists are, in the author's experience, slightly more tolerant to surface oils and greases than dry films with regard to adhesion. Solvents in wet films, especially those with chlorinated solvents, such as KPR-2, tend to penetrate light surface oils. Dry films should be applied only to a surface that has been thoroughly degreased. When one switches from a wet-film process to dry films, it is advisable to check the preapplication cleaning and upgrade it if necessary.

Most of the resists perform better from an adhesion standpoint if laminated while the part is warm. During lamination, the resist is heated when passing over the surface heaters just before it is applied to the part. If the part is cold, it acts as a heat sink and removes heat from the resist quickly. Reduced adhesion will result, although the primary adhesion from lamination is normally adequate for subsequent processing. If the part is warmed, the resist flows and adheres better. A simple test will show this. Laminate a part without heating it, and peel the resist from the surface after about 5 min. Repeat the test with a part heated in a forced convection oven at 180°F for 10 min. After cooling for 5 min, it will be much more difficult to remove the resist from the part if it was properly cleaned in both instances.

Another factor that must be considered when one is cleaning is the physical surface condition of the part before it is coated. Because it is an extremely viscous substance, the resist has some difficulty in flowing into deep surface scratches and defects. Scratches with high aspect ratios are especially bad during etching. The resist will not conform into the scratch but will form a path for the etchant. As a result, a groove will be formed in the conductors, as sketched in Fig. 6-6. The newer resists appear more resistant to these types of defects than the earlier formulations. However, any resists will be affected under some circumstances. The best method of controlling the problem is to prevent it from happening through the use of proper abrasives. An inspection of a scrubbed surface for deep scratches should be made periodically. Abrasives, though rated at a certain level, for example, 4 F pumice, will sometimes contain larger particles, which can cause undesirable, random scratches. Mechanical abrasive scrubbers, such as the Billco unit shown in Chap. 3, will continually circulate these large particles and cause random surface defects to come and go. One method of controlling the problem is to sift the abrasive through wire mesh just before use. The stainless-steel wire that is used for screen

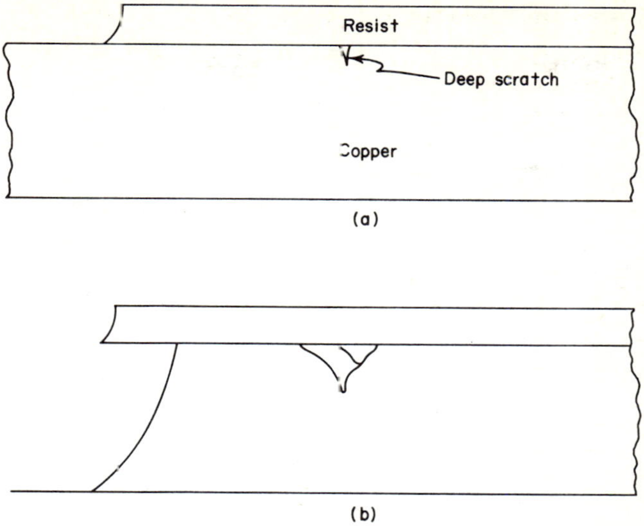

Fig. 6-6 Effect of deep surface scratches. (*a*) Before etching; (*b*) after etching. Defect greatly enlarges to reduce the height of the etched conductor. The etchant enters from outside the photo-resist pattern.

stenciling (140-mesh or better) can be used as a sieve to remove the "golf balls." This type of surface defect is especially a problem when etching fine lines, for example, 10 mils wide in thick copper (3 mils or more).

To summarize, the cleaning process for dry-film resists should contain thorough degreasing steps and should not cause surface defects which the resist cannot completely cover. Preferably, but not essentially, the cleaning process should finish with a light bake to expel moisture from through holes and to heat the part before lamination. Experience indicates that a part should be recleaned if not coated with resist the same day it is cleaned. This generalization is, of course, dependent upon storage conditions; but if there is any doubt, recleaning is advisable. When abrasive scrubbers that do not employ a slurry are used, such as Scotch-brite wheels, a soak cleaner should be used afterward. Table 6-3 shows an example of a cleaning cycle that can be used for dry films. Should parts not be free of water breaks after step 4, a soak cleaner should be employed after the scrubbing and before the acid etch. Anodic cleaning is perhaps the most effective cleaning method for dry films, although racking and unracking time severly limits its value. It preserves the surface condition of the part, degreases thoroughly, and eliminates abrasive scrubbing. Du Pont does not recommend chemical cleaning because of the formation of chemical compounds that could interfere with adhesion-promoter reactions. However, mechanical cleaning is not always a relia-

TABLE 6-3 Example of a Cleaning Process for Dry Films

Step	Remarks
1. Scrub clean: 1 lb 4 F pumice; 1 oz nonionic wetting agent (Triton X-100*); Water to make a slurry	Scrub with a soft bristle brush. Commercial cleaners are also suitable if they contain a detergent. Parts that contain visible oil stains should be degreased before scrubbing.
2. Rinse thoroughly	Remove surface pumice.
3. Etch lightly: 8–12 oz/gal ammonium persulfate; 8–10 vol percent sulfuric acid	Etch 30 to 60 s at room temperature.
4. Rinse thoroughly	Parts should be free of water breaks.
5. Acid dip: 8–10 vol percent sulfuric acid	Immerse 30 to 60 s at room temperature.
6. Rinse thoroughly	Dry parts.
7. Bake: 180 to 200°F for 5 to 15 min	

* Rohm and Haas trademark.

ble degreaser in the author's experience, and chemical cleaning is useful. One surface that resists adhere poorly to is that left by copper pyrophosphate plating when an additive is used. It must be abrasively cleaned.

Lamination

During the lamination operation the photoresist film is applied to the part. Though important, it is not as critical as the resist-application operation during wet-film processing. The one big factor that is eliminated is the resist uniformity problem. As may be recalled, the author gave expected resist uniformity factors and coating thicknesses based upon practical experience for the wet-film coating methods. Dry-film resists have preset coating thicknesses and are very uniform compared with wet films. Thickness variations of ± 0.1 mil are reported, which for a resist such as Riston 30S (2.9 mils nominal thickness) gives a variation of 2.8 to 3.0 mils or a consistent uniformity of

$$\text{Uniformity} = \frac{2.8}{3.0} \times 100$$

$$= 93.3 \text{ percent}$$

For a thin resist 0.75 mil thick, the uniformity is 76.4 percent, which is also a highly acceptable value. In spite of claims to the contrary, few wet-film coating methods will consistently exceed that uniformity. Other preexposure processing is also shortened. Both sides of the part are coated simultaneously, which can be done by only a few coating methods for wet films and those have problems with open-plated through holes. Resist does not flow into the holes and need not therefore be removed during developing.

The actual operation has several steps, all of which are accomplished by a mechanical device referred to as a "laminator" (Fig 6-7). Photoresist films are heated to within a specific temperature range and laminated to parts under pressure for a specific time. The separator sheet of polyolefin is removed just prior to heating. If there is one critical parameter

Fig. 6-7 The Du Pont A laminator (*Riston Products Div. of Du Pont, Wilmington, Del.*)

in the operation, it is the amount of heat that the resist film receives. Most of the operations work to that end. Pressure is also a factor; but if enough heat is available, only a small pressure is required for adequate adhesion.

Factors which affect the heat balance between the part and resist are (1) heater temperature, (2) laminating speed, (3) resist thickness, (4) mass of the part, and (5) temperature of the part. The heater temperature and the laminating speed control the amount of heat transferred to the film by conduction. At thermal equilibrium, the surface temperature of the heater will be approached by the resist surface temperature less thermal resistance. Reducing dwell time of the resist on the heater will reduce the surface temperature of the resist. If the resist is thin, it will become hotter than a thicker resist film with the same dwell time. There is less mass to heat with a thinner resist, and therefore the resist film will approach thermal equilibrium more quickly with the same amount of heat energy. One obvious ramification of this is the increased laminating speed afforded by thinner resists. Another is reducing the laminating speed for thicker resists. However, most resists are usable over a wide latitude of heating, and a speed of 5 ft/min with heater temperature ranging from 220 to 240°F gives adequate results.

The mass and temperature of the part are often overlooked and sometimes important considerations in laminating. Large, heavy parts may tend to function as heat sinks and chill the resist before it has a chance to flow, orient, and adhere. If the temperature of the part is near the temperature of the resist, the problem is eliminated. Many users heat the parts immediately prior to lamination as a precaution, but it is generally unnecessary except for unusually heavy parts, especially with the Riston S series resists and DCR Laminar HS series resists. Heating does help adhesion and it can be demonstrated with a simple test. Using large, heavy parts, such as a piece of brass, heat one to about 150°F and laminate it. Laminate the other part directly at room temperature. When the two parts reach the same temperature, attempt to peel off the sheet of resist in its entirety. Depending upon the resist thickness, the hot laminated part will break into small pieces because the adhesion often will exceed the cohesive strength of the resist. The cold laminated part will generally peel off rather easily. If the part is too hot, the volatile low-molecular-weight adhesion promoters, monomers, and other elements will form blisters in the film. Preexposure baking will not give the same improved adhesion as hot laminating, and it is not recommended as it may cause blistering.

Some hold time after lamination is desirable for any resist, but after the part reaches room temperature, it can be exposed. The 30-min hold time recommended for some resists is more of a precautionary measure. Most of the adhesion-promoting phenomena have taken place when the part reaches ambient temperature.

Some of the resists contain adhesion promoters (37). These are low-molecular-weight substances that are driven away from the heater surface by virtue of their volatile nature. As the part heads in for lamination, these substances condense on the board surface and orient themselves just prior to the resist application. Other low-molecular-weight substances also collect near the copper-resist interface and would appear to contain a high percentage of photosensitive elements.

Exposure

Like other negative resists, dry films are polymerized in areas where light reacts. The molecular weight is drastically increased, which increases the chemical resistance and adhesion to the substrate. One advantage that dry-film resists have is a precise, uniform coating factor. The minimum thickness divided by the maximum thickness expressed as a percent is rarely below 90 percent except on the thinner films. As may be seen from the chapters on wet-film resists, few coating methods approach this consistently.

The discussion will be divided into two parts. Initially, the discussion will consider what goes on during exposure followed by methods of determining and controlling exposure. With reasonably good cleaning, improper exposure is probably the biggest single problem source in dry-film processing. Both aqueous- and solvent-developing films have similar problems. Any factor that tends to reduce or interfere with a correct exposure time can cause problems that may not readily be associated with the exposure operation (Table 6-4). For example, a poor contact in the

TABLE 6-4 Factors Which Tend to Promote Underexposure with an Otherwise Normal Exposure Time

1. Poor vacuum over the part
2. Anything which causes space between the photomask and Mylar overlay
3. Drop in line voltage to the printer
4. Deterioration of the light source (use step tablet to check frequently)
5. Lack of hold time after exposure for the dye-sensitized resists (10 min minimum)
6. Leaks or deterioration of the vacuum system in the printer (check gauge readings frequently)
7. Variations in light intensity at different areas in the printing frame, e.g., corners, edges

vacuum frame or lifting of the Mylar overlay to let air in will give underexposure symptoms with an otherwise correct exposure time. Poor or damaged photomasks will also cause problems, as will dirty printing frames. These types of defects generally cause excessive touch-up and are discussed in detail in that section. However, photomask images with low optical density may force underexposure to obtain any image at all.

Dry-film resists of all types are sensitive to ultraviolet radiation between 3000 and 4000 Å, particularly around 3500 Å. By far, the most common light source is mercury-vapor lamps. Carbon arcs and ultraviolet fluorescent tubes also have a satisfactory spectrum. Light intensity is somewhat important, particularly in tenting holes (Fig. 6-2). When an intense light source is used, polymerization of the film over the open hole is more likely to be successful. Oxygen entrapped in the hole retards the polymerization reactions, but an intense source can generate enough free radicals to consume the absorbed oxygen and cause polymerization to occur faster than fresh oxygen can diffuse into the film.

Oxygen has a significant effect on exposure. As long as the Mylar overlay is present (Fig. 6-1), the film is protected from atmospheric oxygen, and for most applications this is sufficient. Oxygen is present in the photopolymer to some extent, but there is sufficient sensitizer to consume the oxygen and still polymerize the resist.

This initial portion of the exposure, in which the internal oxygen is consumed and the number of activated monomers or other unsaturated components available for cross linking grows to the point where extensive polymerization can take place, is sometimes referred to as the "induction" time, or period.

After the induction period, extensive polymerization takes place in a relatively short time, and large quantities of monomers and other unsaturated components are consumed. During this portion of the exposure, the viscosity of areas in which polymerization is taking place is increased. The higher-molecular-weight polymers created by the exposure form viscous media and thereby become less mobile than the components from which they were formed. Unexposed areas are still loaded with unsaturated lower-molecular-weight monomers and other components. There is therefore a much greater tendency for the less viscous components in the unexposed areas to migrate into the exposed areas than the reverse situation (21). Reactive compounds migrating into the exposed areas will also receive exposure and polymerize if the exposure time is sufficient. Resists that are highly photosensitive and have a high concentration of unsaturated monomers may form a "scooped-out" sidewall with marginal exposure time. Apparently, diffusing monomers are not sufficiently polymerized to endure developing. Straighter sidewalls can be obtained by increasing the exposure time or by heating after exposure to complete the reactions.

Polymerization continues until the energy available is consumed or until the concentration of unsaturated components reaches a low level. A plot of monomer consumption versus exposure time from several sources (22,23) with the three zones just discussed superimposed is shown in Fig. 6-8. If the total exposure energy is not sufficient to complete the induction period (Zone I) and start the rapid consumption of unsaturated compounds (Zone II), experience has shown that no useful image will be obtained because of extensive damage during developing. In other words, the difference between the exposed image and the nonimage areas is not enough to prevent extensive penetration and dissolution or damage by the developing solution.

In Zone II, where extensive polymerization starts, lie the most accurate exposure times. Accuracy refers to the minimum deviation of the resist image from the photomask. Figure 6-9 shows a plot of line-width deviation from a photomask versus the exposure time (24). For this particular

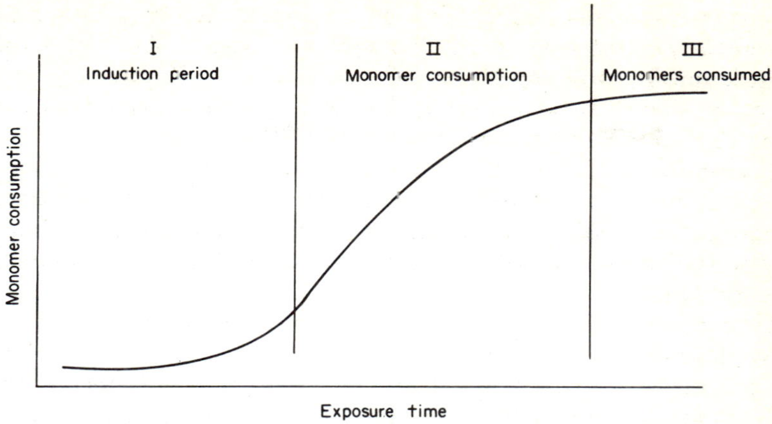

Fig. 6-8 Monomer consumption versus exposure time (22,23).

resist thickness and light source, image deviations of over 1 mil can be obtained with extensive overexposure. However, the condition of the resist image is also improved slightly by the longer exposure time and is thereby less affected by the factors which promote underexposure (Table 6-4). Since the adhesion of the photoresist is dependent upon polymerization, at least an adequate exposure becomes extremely important, particularly for plating applications.

The last zone (III) occurs when the available monomers and unsaturated elements are essentially consumed. Exposure well past this point will tend to promote scum in the unexposed areas from heat (heat fog).

Hold time after exposure is important for some types of resists. This

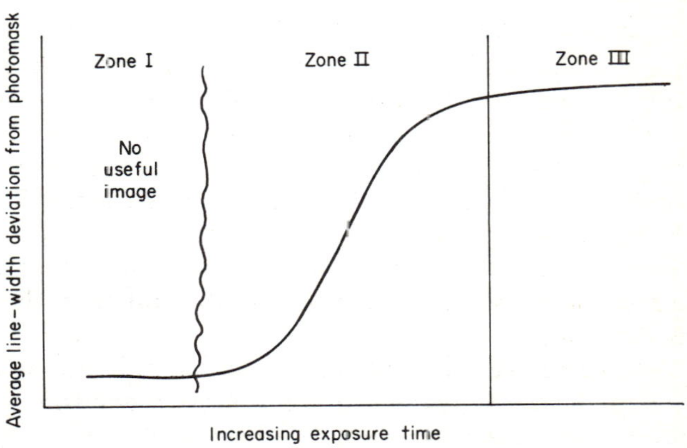

Fig. 6-9 Line-width variation from photomask with increasing exposure time.

can be determined by exposing five panels coated with the same type of resist using a step tablet for a photomask. Use the same amount of exposure time and the same printer. Immediately after exposure, develop one panel. Wait 5, 10, 15, and 30 min for the next four panels before developing. Depending upon the type of resist and step tablet, there may be a difference until 10 or 15 min has elapsed.

Heat is also a factor during exposure. The polymerizing reactions that are initiated by light can be continued by heat energy. Printing frames can sometimes grow quite warm, particularly after continued use with high-pressure mercury-vapor lamps. Heating tends to cause overexposure when using an otherwise adequate time. Probable results include scum in nonimage areas and some line-width growth. One method of improving accuracy when necessary is to expose the image just slightly greater than the induction (Zone I) time and to heat to complete the reactions well into the second zone. Resists can be baked at 180°F. Times will vary with the mass of the part but should be enough to ensure that it reaches the temperature for a short time (about 5 min). Hold times are generally unnecessary when using this technique, and other processing is unchanged. However, the author's experience with this method has not been altogether satisfactory.

There are two basic methods of determining exposure time, depending upon the image accuracy required. For general use, reasonable accuracy can be obtained using step tablets. Where fine lines and spaces are involved, one uses line-width measurements to set an exposure time followed by a step tablet or resolution pattern to fix the exposure with a visual reference. The other methods mentioned in the Chap. 4 section on exposure are also generally applicable, but discussions herein will center on the two aforementioned methods.

An important thing to remember about the step tablets is that they give only one measured parameter. The number of the threshold step is a visual indication of the quantity of light reacting with the photoresist. The step tablets (see Chap. 4) irradiated with actinic light will transmit varying quantities of light through each step because each step has varying optical density. Assuming that the power of radiating energy (light) is constant over the entire area of the tablet, the amount of light energy per unit area is then directly proportional to the time. However, the exposure time and the power per unit area do not vary over the entire step tablet. What does vary is the power transmitted by each step. Steps exposed through the tablet into the photoresist will not become sufficiently polymerized to endure developing unless a minimum amount of light energy is absorbed and reacted, assuming that the resist thickness is constant. The clearest step in the tablet will transmit the most power to the photoresist beneath and the darkest step the least. When the light energy received by step 6 or 7[1] is sufficient to render either or both steps insoluble, the areas of resist outside the step tablet have received the proper level of light energy to form an accurate, adequately polymerized image. Dye-sensitized resists perform best at a threshold step of 8 or 9, but often reach this level with less exposure time because of greater reactivity with light energy.

It is apparent that the condition of the threshold step does not reflect the image condition for reasons just explained, and no such interpretation should be made. If the threshold step is low, for example, 3 or 4, exposure time should be increased, even if the photomask image exposed alongside the step tablet looks good. Chances are that unnecessary losses will be experienced in subsequent processes because of poor adhesion and resist degradation. Overexposure can cause brittleness, image growth, and, possibly, some loss of adhesion. Generally, however, it is the better of two evils.

Using a photomask, or a portion thereof, alongside the step tablet is helpful in detecting potential problems. It is true indication of what the final circuitry will look like at the exposure level indicated by the threshold step. The test panel can actually be etched or plated if desired. Line widths can be measured or examined under magnification. The step tablet can also be placed between the photomask and the resist if readings thus obtained are then related to line-width measurements and image condition. Care must be taken to ensure good contact between the mask, step tablet, and resist layers.

Variations in the optical density of the steps occur in a predictable manner. It is therefore possible to predict the exposure time required to obtain a given step from the results of an arbitrary exposure time. This can be done through the use of equations or tablets, as shown in Chap. 4. There are several types of step tablets in use. Stouffer, Kodak, and Du Pont all make several kinds of step tablets. The most important consideration is the incremental step density. The most commonly used value is the 0.15-increment 21-step tablet discussed in Chap. 4. With an incremental density of 0.10, thirty steps are used and with 0.05, forty-one steps. Recently, Du Pont made available a 17-step tablet with a 0.05 incremental density specifically for Riston. In general, the larger the increment, the easier the tablet is to use, but the less the accuracy. For example, the clear portion of a 21-step tablet has a density of only 0.05, up to 3.05 for the last step. It is virtually impossible to expose long enough to insolubilize all the steps and rarely is the exposure short enough not to

[1] 21-step 0.15-density gradient, e.g., Kodak No. 2.

insolubilize a threshold step. As far as accuracy goes, however, the range covered by steps 6 and 7 would cover steps 8 to 9 on the 30-step tablet and steps 16 to 19 on the 41-step tablet (Table 6-5).

TABLE 6-5 Relationship between the Density and Step of Various Step Tablets

0.15 density 21 steps	0.10 density 30 steps	0.05 density 41 steps	
1. 0.05	1. 0.05	1. 0.05	19. 0.95
2. 0.20	2. 0.15	2. 0.10	20. 1.00
3. 0.35	3. 0.25	3. 0.15	21. 1.05
4. 0.50	4. 0.35	4. 0.20	22. 1.10
5. 0.65	5. 0.45	5. 0.25	23. 1.15
6. 0.80	6. 0.55	6. 0.30	24. 1.20
7. 0.95	7. 0.65	7. 0.35	25. 1.25
8. 1.10	8. 0.75	8. 0.40	26. 1.30
9. 1.25	9. 0.85	9. 0.45	27. 1.35
10. 1.40	10. 0.95	10. 0.50	28. 1.40
11. 1.55	11. 1.05	11. 0.55	29. 1.45
12. 1.70	12. 1.15	12. 0.60	30. 1.50
13. 1.85	13. 1.25	13. 0.65	31. 1.55
14. 2.00	14. 1.35	14. 0.70	32. 1.60
15. 2.15	15. 1.45	15. 0.75	33. 1.65
16. 2.30	16. 1.55	16. 0.80	34. 1.70
17. 2.45	17. 1.65	17. 0.85	35. 1.75
18. 2.60	18. 1.75	18. 0.90	36. 1.80

To calculate the proper step from a known exposure time and the resulting step, the equation and procedure in Fig. 6-10 can be used. An easier method is to use the data in Table 6-6. For example, if an exposure time

$$\text{Exposure time for the desired step} = \text{Reference time}\left[\sqrt{2}\right]^{(\text{Desired step} - \text{reference step})}$$

$$t = t_R\left[\sqrt{2}\right]^{(S - S_R)}$$

Formula for calculation of exposure times when using a 21-step 0.15 step density tablet

Fig. 6-10 Equation used to calculate desired exposure time from the reference time and step of any previous exposure.

TABLE 6-6 Multiples of Exposure Time to Obtain Steps 6 and 8 on a Kodak No. 2 Step Tablet

Step* obtained by exposure	Multiple to obtain step 6	Multiple to obtain step 8
1	5.66	11.13
2	4.00	8.00
3	2.83	5.66
4	2.00	4.00
5	1.41	2.83
6	1.00	2.00
7	0.71	1.41
8	0.50	1.00
9	0.35	0.71
10	0.25	0.50
11	0.18	0.35
12	0.13	0.25
13	0.09	0.18
14	0.06	0.13

* First shaded step, regardless of number, on the tablet.

of 30 s gives a threshold step of 3 for a dye-sensitized resist (one that gives a darkened latent image after exposure), the proper exposure time is

$$T = Kt$$

where

T = proper time to obtain a threshold step of 8

t = time to obtain a threshold step of 3, which is 30 s ($\frac{1}{2}$ min)

K = value from Table 6-6 at a step of 3

or

$$T = (5.66)(30)$$
$$= 2.83 \text{ min, or } 2 \text{ min } 50 \text{ s}$$

Under the same conditions, a non-dye-sensitized resist requiring a step of 6 has a time of about 1.42 min, or 1 min 25 s. However, the data in Table 6-6 are useful only in the 0.15-density 21-step tablets.

It is also possible to simply bracket the proper exposure time by a series of exposures with a step tablet. To begin with, several panels are cleaned and coated with the resist. After the panels have cooled, the step tablet and a small portion of a photomask are placed side by side on the panel in the vacuum frame. For printers that employ a fixed light source and a movable vacuum frame, an initial setting of 1 min will usually be sufficient to obtain a step to indicate whether to increase or decrease the time.

The condition of the image formed from the photomask should also be noted in comparison with the threshold step. Depending upon the indications of the initial exposure, the time is increased or decreased stepwise until the right exposure is obtained.

The threshold step may not always be clear-cut. Generally, if more than one-half of the step survives developing, it should be taken as the threshold step.

Scanning printers in which the vacuum frame is fixed and the light source moves are a different situation. The dial indication is not linear with time like the timing clock on the fixed-light printers. To use the equations or tables, the settings must be referenced to a dwell time over a set distance, for example, 1 ft. Table 6-7 shows an example. If a setting of 50 gives a step of 6 on the scanning printer of Table 6-7, which is a reference time of 16 s, the proper time could be estimated as follows for a dye-sensitized resist:

TABLE 6-7 Reference Times and Settings for a Scanning Printer

Setting	Time to cover 12 in. (s)
0	Stall
10	Near stall
20	45
30	27
40	22
50	16
60	13
70	10
80	7
90	6
100	5

$$T = Kt$$

$$= 2.00(16) \text{ (value to go from step 6 to 8 from Table 6-6)}$$

$$= 32 \text{ s}$$

From the plot of the data in Fig. 6-11, the actual setting is 24 to obtain a step of 8.

The second method involves actually measuring the line widths result-

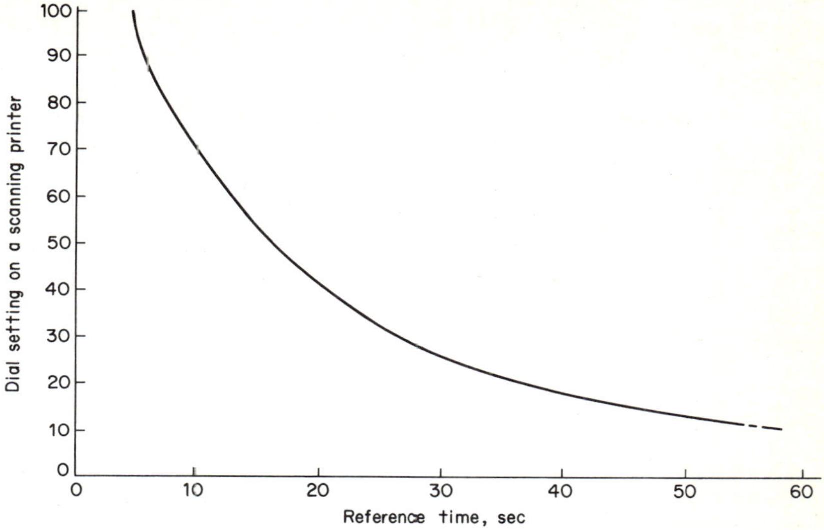

Fig. 6-11 Plot of scanning-printer data (Table 6-7).

ing from a given exposure time. Points are measured on a photomask and relocated on the resist image and measured. By adjusting the exposure time, the measured difference is made as small as practical. Table 6-8 shows the results of line-width differences with varying exposure time. Seven panels were coated, exposed for the time shown, and developed. Using the overlay system shown in Fig. 6-12 to locate points on the resist

TABLE 6-8 Variation of Line-Width Accuracy with Exposure Time

Exposure time (min)	Average line-width deviation from photomask (mils)	Range of measurements (mils)
1.0	No image	
1.75	Nil	+0.2 to −0.1
2.0	Nil	+0.2 to −0.2
2.5	+0.42	+0.2 to +0.8
3.0	+0.71	+0.5 to 1.1
4.0	+0.81	+0.6 to 1.0
5.0	+1.01	+0.7 to 1.4

NOTES: (1) Each exposure time had 12 measured points on a single part. (2) Light intensity was 7 mW/cm^2. (3) Resist was Riston 15 exposed with a fixed 1,000-Watt mercury-vapor lamp at 3 ft.

image that were measured on the phototool, 12 points were measured for each exposure. An average was taken (Table 6-8) and plotted in Fig. 6-13. Using a maximum difference of 0.3 mil or less gives an exposure time of 2.33 min or 2 min 20 s. This would be the preferable time. To verify the time, four additional parts were exposed at 2 min 20 s. Of the 48 measurements, the range was from +0.1 to −0.3 mil, with an average of −0.1 mil. A step tablet revealed a solid 7 was held on a Kodak No. 2.

1. Prepare clear semiridged plastic overlay registered to photomask. Locate specific points on the circuitry by marking. Separate the overlay and punch small holes at datum points. Reassemble and measure line widths through points. At least five points per side should be measured.
2. Use the measured photomask to conduct a series of identical exposures except for varying the exposure time. Develop and etch lightly (or completely) to check for scumming and to delineate the circuitry clearly.

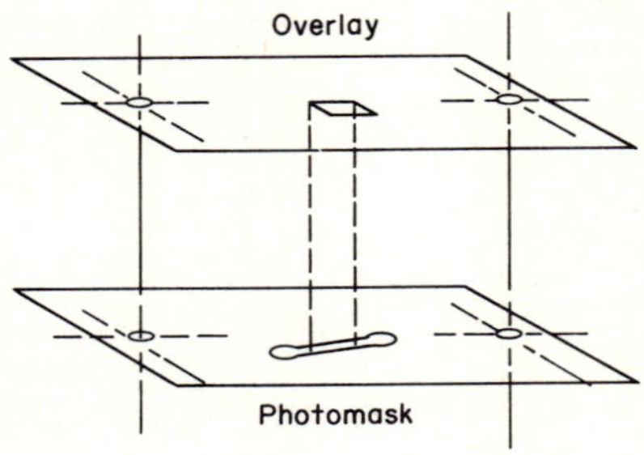

3. Align the plastic overlay to the photoresist pattern and measure the resist image. Compare the line-width measurement of the resist image to the photomask and note the deviation. Average and plot deviations (Fig. 6-9) to obtain the most accurate time. Measure lines at base.

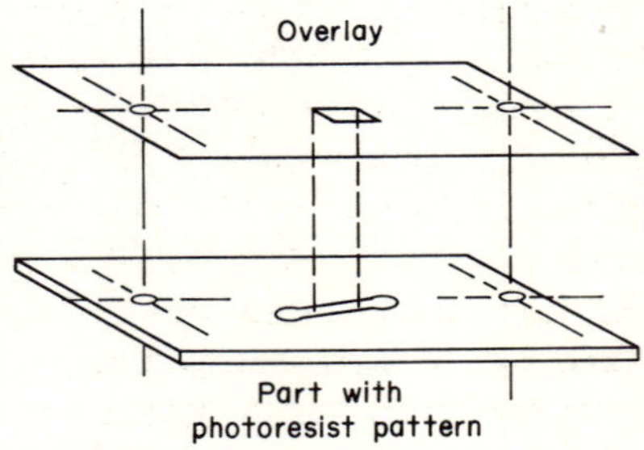

Fig. 6-12 Exposure-time determination by line-width measurement.

Using a different image, four parts were processed with similar results. For production purposes, the step-tablet setting was used with periodic remeasurements.

Sometimes, a light source may not be sufficiently collimated to give an accurate image. This fact is indicated by a low-step-tablet value for an average difference of ±0.3 mil or more. Obviously, if an acceptable value cannot be obtained, collimation is not good, although the method of measurement must be accurate with good sidewall configuration on the resist.

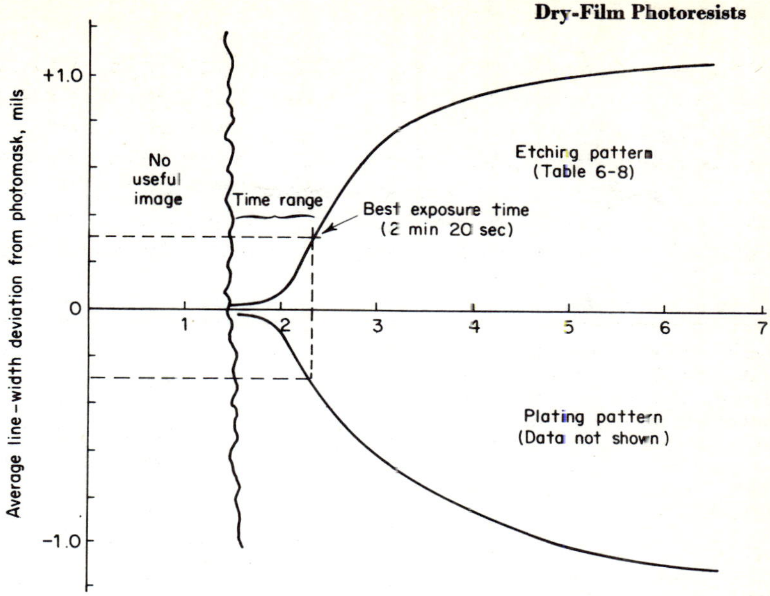

Fig. 6-13 Exposure time and line-width deviation.

Developing

As with the other negative resists, developing removes the unexposed portions of the dry film to form the image. There are some similarities between aqueous- and solvent-developing resists, and they will be discussed together except as noted. Two types of processors will be discussed, the Du Pont C Processor (Fig. 6-14) and the A Processor (Fig. 6-15). While there are aqueous processors made of polypropylene and PVC or entirely of PVC, such as converted chemical etchers, there is enough similarity to the C Processor to adapt the technique. As with any conveyerized system, it is necessary to relate the conveyer setting to the dwell time in the active zone (spray chamber) or to the result of processing (over- or under-developing).

The majority of the discussion will center on solvent developing because of the higher usage rate.

Determining Developing Times with the C Processor. The dwell time in the C Processor developing and water-rinse chambers ranges from about 30 to 105 s over the 6-ft length. This corresponds to a rate of about 3.4 to 13.0 ft/min. Actual ranges may vary lightly from machine to machine. Conveyer settings have a range of 1 to 10, with the former near stalling and the later near or at maximum speed. The relationship between conveyer setting and dwell times should be determined by timing a part from its penetration into the first spray chamber until it leaves the water rinse for each of several settings. A suggested range of conveyer

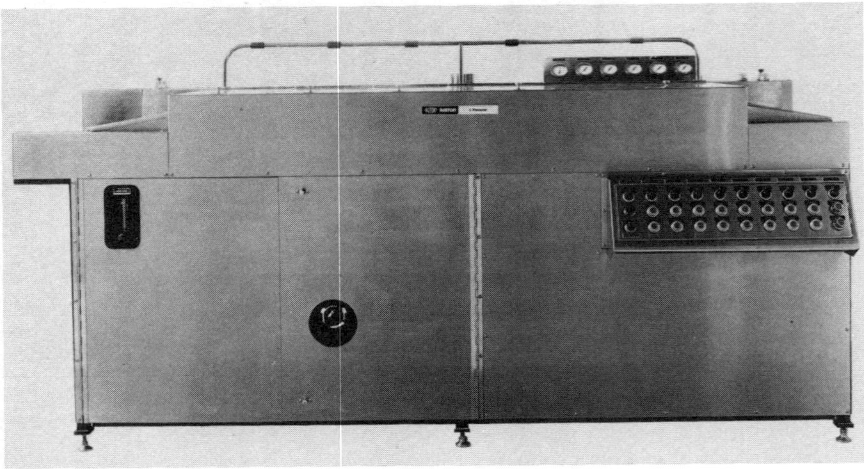

Fig. 6-14 The Du Pont C Processor. (*Riston Products Div. of Du Pont, Wilmington, Del.*)

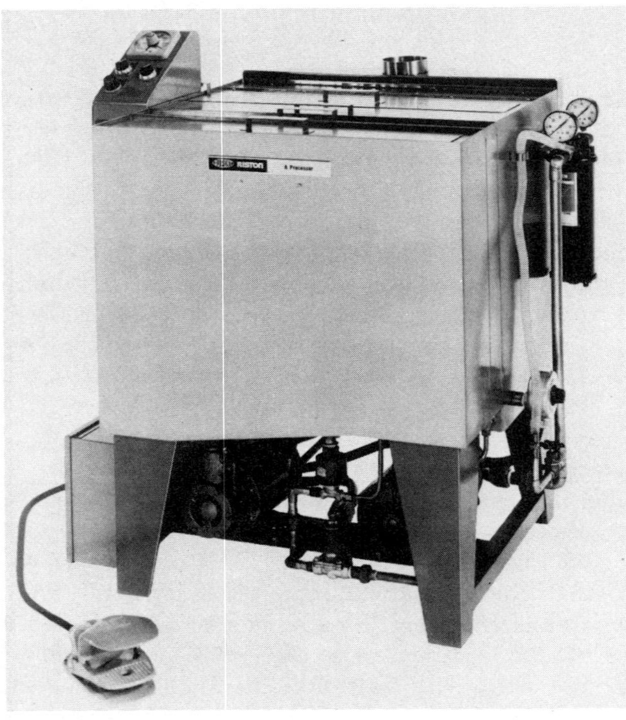

Fig. 6-15 Du Pont A Processor. (*Riston Products Div. of Du Pont, Wilmington, Del.*)

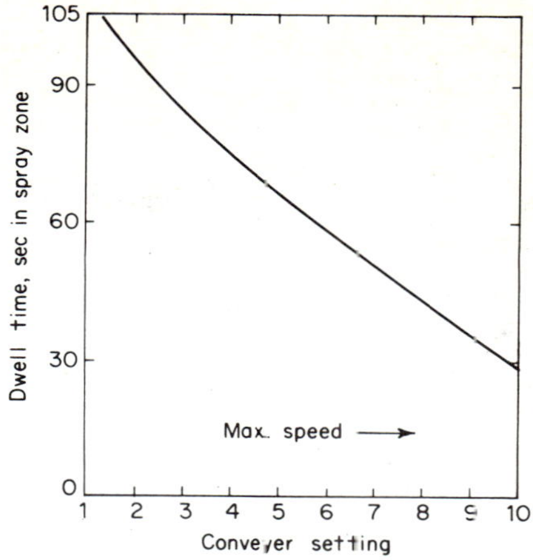

Fig. 6-16 Example of conveyer setting–dwell time plot for a C Processor.

settings is 1.5, 3.0, 5.0, 8.0, and 10.0. Use a typical part because there may be a slight variation between thin, flimsy parts and large, heavy parts. A typical plot is shown in Fig. 6-16 from Table 6-9, but one should prepare one's own. In Fig. 6-17, a plot of dwell time for various thicknesses of resist is shown. These settings can be used as starting points in determining actual developer settings. The dwell time indicated should be bracketed by one unit on the dial setting in either direction except for the very thick ($\approx$3 mils) or very thin ($<$1.1 mils). For example, if the film thickness of solvent-developing resists is 2 mils, a dwell time of about 70 s is indicated,

TABLE 6-9 Relationship between Conveyer Setting and Dwell Time

Conveyer setting	Dwell time
1.5	1.72 min (3.5 ft/min) 103 s
3.0	1.40 min (4.3 ft/min) 84 s
5.0	1.09 min (5.5 ft/min) 65 s
8.0	0.73 min (8.2 ft/min) 44 s
10.0	0.46 min (13.0 ft/min) 28 s

which would require a dial setting of 4.5 units with tests at 3.5 and 5.5. The test parts should be exposed properly and the rinse solvent clean. Very thick films, such as 3 mils, approach stall speeds: at 3.0 mils, 105 s, or somewhere between stall and 1.5 on the conveyer dial. Under these conditions, only an increase in conveyer setting can be evaluated. The same is true when operating near the maximum speed.

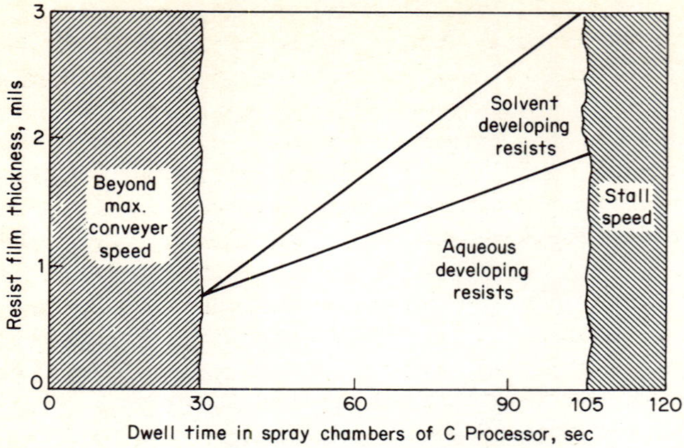

Fig. 6-17 Starting points to determine conveyer settings on C Processors.

When developing multiple layers of resist or specialty resists, such as Riston 50, the necessity of multiple passes through the processor arises. For 5 mils of resist, a pass at about 1.5 (3 mils) without water rinsing followed by a second pass at 4.5 (2 mils) with the rinse is usually adequate. It is advisable to invert the part between passes and to transfer between passes as rapidly as possible.

Aqueous developing is similar to solvent developing, except that the resists are somewhat slower and the solution temperature is higher.

Temperature Control. During solvent developing, the temperature of the developer is important and should be within 5°F of 65°F. If the temperature gets much below 60°F, incomplete developing is probable due to reduced solubility and dissolution rate of the resist in the solvent. An increased flow rate and longer developing times will be required to avoid scum, if it can be avoided. Much above 70°F, the activity of the solvent is too great, and unnecessary resist damage is possible. Here, again, the result will be scum. Damaging the resist may cause liquid plasticizing agents to bleed out onto the metal and cause scumming. In extreme cases, the image will be lost or damaged, particularly if it has received incomplete or marginal exposure. Exposure tends to chemically tie up the plasticizing agent, which reduces bleed out. Volatile liquids, such as the methyl chloroform developer, will tend to heat up from the work done on the solution during pumping. Without chilling or extensive cooling from evaporation, the solvent temperature may reach 90°F or more. Conversely, if the processor is not operated and the cooling unit left on, temperatures may drop below 60°F. Best results are obtained by achieving an equilibrium temperature of 65°F with the pumps and cooling unit operating and being left on during work periods.

Aqueous-developing resists can tolerate, and actually need, a higher developing temperature. A temperature around 80°F will develop most aqueous resists satisfactorily. Depending upon several factors, such as the type of spray equipment, the duration and frequency of pump operation, ventilation, etc., temperature controls may not be necessary. Most converted etchers have cooling coils and heaters to maintain a constant temperature. One advantage of aqueous type of resist processing is the high capacity of the developers; 5 to 15 ft^2 of circuitry per gallon is readily attainable.

Other Factors. The C Processor contains filters which must be cleaned regularly to ensure a full flow of solvent through the nozzles for proper developing. It is actually a good idea to inspect them once each shift if the processor has been worked hard, but, at the least, some inspection routine should be established. Sometimes, parts may contain small alignment patterns in the middle of through holes on boards to be pattern-plated, e.g., crosses, squares, etc. During exposure, the small areas are hardened. The pad area and the area over the hole are not exposed during pattern plating and therefore are dissolved. The small, unattached images thus created tend to tack on to hole walls and image areas. So do brittle flakes at the board edges that have been hardened by exposure and break off during developing. Bits of resist that pass through the filter will clog nozzles and give uneven spray patterns, which contribute to scum. Periodic inspection of the spray pattern is desirable.

The flow rate of fresh developer into the processor is also important. Developer enters the processor via the last chamber, cascades into the next chamber upon demand, and cascades finally into the initial chamber. Spent developer overflows into a scrap-solvent tank or exits direct. When too much resist is processed for the developer flow rate, parts that are not nearly developed will reach the last chamber. While the developing time has not changed, the effectivity of the developer has, because it is approaching saturation in the first chamber. As a result, the last, or clean, chamber starts to become contaminated. Some contamination can be tolerated here; but as soon as enough resist is accumulated, scum will occur as the resist-laden solvent enters the water-spray rinse chamber. Initially, scum will be noted only on one (top) side but eventually on both sides.

This type of scumming can be avoided by periodically checking the last developer chamber for resist dye. There should be none. If there is, the flow rate is not presently adequate for the parts processing rate. Obviously, a flow rate can be adequate for long runs, but can still be overloaded temporarily.

Developer-spray pressure gauges should be noted. At least 15 psig should be maintained, depending upon the type of spray nozzles.

Low water pressure can also promote scumming, particularly on the top of the part. Pressurized water spray literally blasts the resist-laden solvent from the part. If the pressure is too low, water can precipitate the resist from the solvent that is not physically removed. Pressures as low as 12 psig can be used, but fine-line, dense patterns will generally require at least 15 psig, preferably 20 psig.

Developing with the A Processor. The A Processor, a moderate-volume unit, gives good results when properly used. Because of the low developer capacity, there is a tendency to use the developer too long before going to the trouble of dumping. This is a source of problems which will appear in the form of scum or, in extreme cases, incomplete developing. Another condition that can contribute to scumming problems is the transfer times and the operator dependence thereon. If too long is allowed between the spray and the clean developer rinse, the resist may be redeposited from the drying solvent. Although the film is still soluble, allotted time in the soak tank may not be adequate to completely remove it before water rinsing. Generally, poor results will occur if the developer is allowed to dry before the water rinse. If the developer has evaporated, water will often not remove the film of redeposited resist and can actually make it harder to remove upon redeveloping.

During the recommended operational sequence, the first chamber used is the spray. For all practical purposes, nearly all the developing takes place there. Racked parts are placed in the chamber, the lid closed, and the timer started which starts the spray. The low capacity of the chamber (7 gal or less) makes the square footage allowable small, particularly for thick resists. When the solvent becomes contaminated with resist resin, the developing time may become longer. Also, when individual spray nozzles clog, uneven spray patterns can result, which will cause variations in the developing pattern. This can easily be interpreted as scumming. Increasing the developing time will damage the areas already developed and may cause plasticizer bleed out. There are also slight variations in the spray patterns in the processor, particularly in the extremities of the rack. One method of reducing this problem is to set the timer for about half the normal developing time and to turn the rack 180° before finishing the developing. Solvent evaporation may reduce the volume and concentrate the dissolved resist.

Following the spray developing, an immersion in fresh developer is used to dilute the resist-laden developer carried in from the spray to a very low level. Only a short time and mild agitation are necessary to complete this step. In fact, care should be exercised to keep this time to minimum (5 to 10 s) to avoid resist damage. Developer in this tank should be kept clean (and dry if it is methyl chloroform) to minimize developing problems with the A Processor. If it becomes highly contaminated with dissolved resist, redeposition of resist onto the part will often occur in the subsequent water rinsing. High water content in this tank when developing with solvents will also cause scumming problems. The water level that causes problems with methyl chloroform is not known but may well be around 300 ppm or more. When the developer in the spray tank becomes contaminated, it can be replaced by the relatively clean developer in the immersion tank. Fresh developer may then be added to replace the transferred developer, thereby ensuring a clean immersion tank.

The final operation contained in this A Processor is the water rinse, which is very important. It literally blasts the resist-laden developer from the part; this prevents drying and redeposition. Low water pressure, uneven spray patterns, or high water temperature (80°F or more) can all promote scum formation by redeposition when using solvents, but generally has no ill effects on the aqueous-developing resists. Figure 6-18 shows spray developing time as a function of resist thickness for solvent and aqueous resists. These curves are set as starting points in determining the proper developing time.

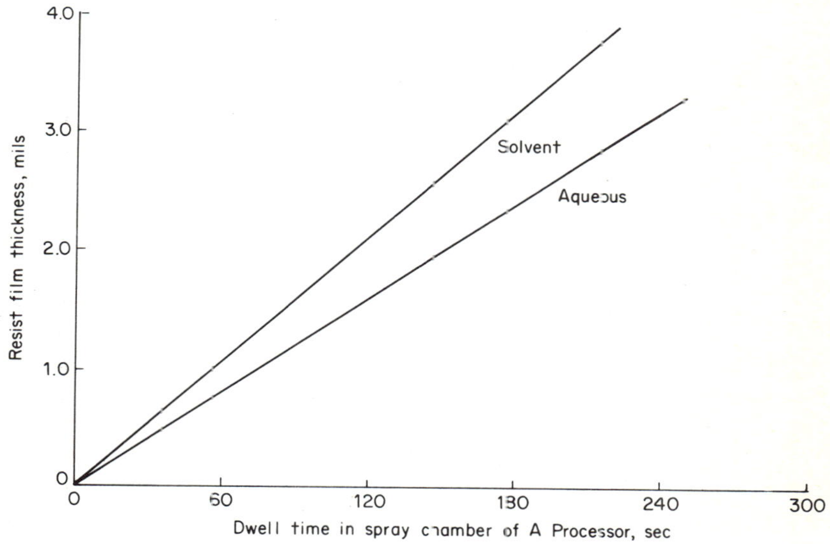

Fig. 6-18 A Processor developing times.

The A Processor is particularly well suited for the aqueous resists. The very high capacity of the developing solution, its low volatility, and its moderate temperature requirements are well suited for this unit.

Scum. If developing problems could be described in one word, it would be "scum." Such is a thin organic film in the nonimage areas that is created during the developing operation. Four basic sources of scum are:

1. Incompletely removed photoresist
2. Resist constituents that are precipitated from the developing solution onto the part
3. Resist constituents that bleed out during and after developing
4. Films formed by non-resist-related contaminants precipitating from the developer

Some scum is frequently present but normally causes no processing problems. Etchants easily penetrate it, and preplating cleaning solutions remove it readily. This is the type mentioned initially. However, it can form heavier films that do cause subsequent processing problems when anything causes incomplete developing, such as too short a developing time or too high a concentration of resist in the solvent. The second type of scumming is caused by anything that reduces the solubility of resist constituents that have been dissolved in the developing solvents. Primary examples include a high level of water contamination in the developer or a poor water rinse. If the resist image is overdeveloped or underexposed and developed for normal times, liquid plasticizing components may bleed out into the nonimage areas. This is the third type of scumming. It can also occur with some resists on parts that have been developed and allowed to set for several days before plating or etching. The fourth type of scumming is rare. Most often, it is the result of the developer solvent coming in contact with incompatible plastic materials. Plasticizers are sometimes leached from incompatible plastic pipes, or there is some related incident. Whether it interferes with resist solubility and prevents complete developing or actually precipitates on the part itself is not known. While rare, it can occur. Sometimes, developing different types of resist together can also cause scumming.

A very effective method of reducing scumming problems during plating is to include an anodic cleaning step in the preplating operations. Scum can also be readily detected prior to etching or plating by treating with a contrast solution. The solution reacts with copper to form red cuprous oxide. Fine scum is readily removed by the solution. Clean copper beneath heavy scum retains its characteristic color, which contrasts sharply with the reacted copper and the resist image. The contrast film dissolves readily in most etchants and does not normally need to be removed. About 0.1 mil of copper is removed by the treatment, which consists of immersing it for 1 to 2 min at ambient temperature in the following:

6.7 oz/gal cupric chloride dihydrate ($CuCl_2 \cdot 2H_2O$)
20 vol percent hydrochloric acid

Mild agitation will give greater uniformity. After sitting for some time, the reacted copper surface will turn from red to black with the formation of cupric oxide, particulary when exposed to light.

Solvent Developers. The developing solvent is methyl chloroform, or 1,1,1-trichloroethane, as one may prefer to call it. Although the chemical name given to the solvent infers a single compound, it is not pure. Chemically, it is too reactive with common metals, such as mild steel, aluminum, and zinc, to be sold in the pure form. Corrosion inhibitors are invariably present, which often affect the developing action of the solvent. An example is dioxane (25), which is effective in a concentration of about 4 percent to reduce corrosion of aluminum and black iron pipes. Dioxane, prepared from ethanol, is a very effective solvent for many dry-film resists. Aliphatic nitrate compounds are also useful to reduce corrosion (26) and are present, but in much lower concentrations, about ½ percent if patent literature is any indication.

Lower aliphatic chains containing an epoxide segment are effective at reducing the corrosion of zinc in the vapor phase (27) in a concentration of 1 percent or less. An example is butylene oxide. The epoxide reacts readily with any hydrochloric acid formed by the reduction of the trichlorinated methyl group in the solvent molecule and thereby prevents corrosion. Various amines and pyrroles may also be present.

The solvent is available in two basic formulations, cold-cleaning and vapor-degreasing grades. As a general rule, the vapor-degreasing grade contains more inhibitor and will be a more reactive developer. If it is used, the temperature and developing time must be more carefully controlled. Conversely, there is less scumming tendency. The cold-cleaning grade is safer over a wider temperature range and is generally specified by dry-film resist manufacturers. In some instances, the difference is not significant with regard to dry-film developing. From patent literature,[1] a solution of methyl chloroform contains approximately the following:

1,4-Dioxane	4 percent
Nitromethane	0.5 percent
1,2-Butylene oxide	0.5 percent

In the patents, no distinction was made as to whether the inhibitors would be specific for degreasing, except for the butylene oxide. One might thereby surmise that it would be present in a higher concentration in the degreasing grade than the cold cleaner.

If one point can be made, it is to be careful in selecting a solvent for developing and even more careful in changing solvents. Find a suitable solvent that works, and stay with it. If it becomes necessary to change evaluate the flow rate required to maintain good parts. Water content of the solvent is a factor. Scum will often occur when the water content reaches 300 ppm or more.

[1] Patents assigned to Dow Chemical (25–27).

Some users prefer reclaimed solvent for economy. However, care should be exercised. There are presently at least three other large manufacturers of methyl chloroform besides Dow Chemical. At least two appear to use inhibitor systems that are similar to one previously described. Another appears to differ significantly. Also, the raw-material sources can vary in purity. Typically, compounds such as ethylene dichloride and vinyl chloride may be present as a result of manufacturing in all brands, but the amounts differ. All the fresh solvents are used successfully to develop the resists, but when mixed together and reclaimed problems will invariably occur. A generally satisfactory compromise can be reached by mixing new and reclaimed solvents. Start with a fresh drum of solvent; then every other drum can be reclaimed solvent, and one can still get generally good results. Stills, such as those shown in Figs. 6-19 and 6-20, can be used without some of the problems of reclaimed solvent. There is, of course, no danger of unknown solvent inhibitors becoming mixed because the user controls the input. Generally, somewhere between 70 and 90 percent of the solvents can reportedly be recovered and recycled using stills.

Fig. 6-19 Solvent recovery still for use with the C Processor. (*Riston Products Div. of Du Pont, Wilmington, Del.*)

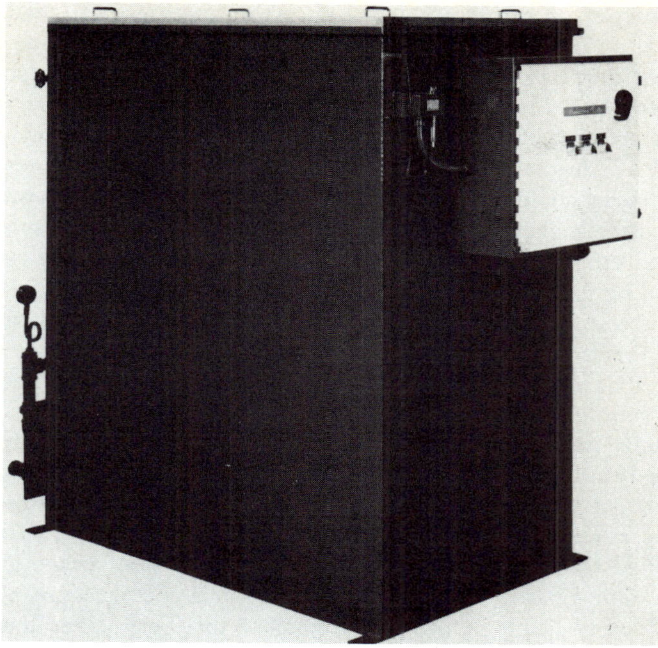

Fig. 6-20 Solvent recovery still for use with the conveyerized spray stripper. (See Fig. 6-21.) (*Riston Products Div. of Du Pont, Wilmington, Del.*)

Aqueous-developing Solvents. Aqueous developers contain one essential ingredient—a source of alkalinity. Dynachem employs sodium carbonate or trisodium phosphate in low (2 wt percent) concentrations The Du Pont developer 2000 also contains sodium ions. Type II resist resins characteristically contain carboxylic acid groups, which will react with strong bases to form a soluble salt (23):

$$\text{(Resin) COOH} + \text{Na}^+ + \text{OH}^- \longrightarrow \text{(Resin) COONa} + \text{H}_2\text{O}$$

Obviously, all the resin cannot be prepared wholly from acid monomers without problems. If too many are present, the exposed areas could be damaged by the developing cycle, or adhesion problems could occur. Too few would, of course, make the system too difficult to develop in aqueous solutions.

In addition to a source of alkalinity, foam-suppressing chemicals may be added or may be present in proprietary formulations. Resist constituents tend to foam when agitated after being neutralized with alkalies, not unlike soaps and anionic detergents. Common foam suppressors include long-chain aliphatic alcohols. Foam-suppressing chemicals are available from both Du Pont and Dynachem for their respective products.

Du Pont developer contains a liquid-solvent constituent in addition to the dry powder. From the odor, one would guess that it contains ethylene glycol monobutyl ether, a compound which is better known as butyl Cellosolve. Alles mentions the use of this and other solvents in his patents (15,17) on aqueous-developing resists to dissolve some resist constituents. A less volatile higher-molecular-weight solvent, such as diethylene glycol monobutyl ether, could also be used. Isopropanol was also mentioned, but would be too volatile for spray developing.

There are several sources of alkalinity that could be used in addition to the phosphates and carbonates used to develop the Laminar A resists. Borax at 10 g/l concentration was mentioned. Satisfactory sources also included sodium metasilicate and triethanol amine.

Generally, the makeup of the developers is not extremely critical. Adequate results have been obtained with the Dynachem Laminar with 1.5 to 3 oz/gal sodium carbonate. Du Pont Developer 2000 requires 10 to 13 vol percent solvent and about 1.3 oz/gal dry salt. Makeup kits are supplied that virtually eliminate any measurement other than water volume. One gallon of solvent concentrate and one measure of dry salt can be added to 7 to 9 gal of water to make 8 to 10 gal of developer. Tap water is adequate unless a precipitate is noted when mixing.

Too high a concentration of caustic often will retard rather than increase the developing speed, contrary to what one would deduce. The reason is not apparent, but possibly the solubility of the reacted resin salts is decreased at high caustic concentrations.

Touch-Up

Touch-up was drastically reduced, even with difficult parts, by the introduction of dry-film resists. In most instances, however, it still exists. There are two problems that are associated with the touch-up of any photoresist image. These are the replacement of missing photoresist in the image area by an organic mask-out and the removal of extraneous photoresist remaining in the nonimage areas.

To replace missing portions of the image with another organic material, it is necessary that the solvents in the touch-up material not extensively damage the photoresist. It is also desirable that the ink be strippable in the same solvents as the photoresist. This is especially true of the resists that are to be stripped in aqueous media. The ink must also adhere well to the copper and photoresist. Screening inks do well as touch-up inks generally. In addition, these inks are formulated to have a long pot life and are available in either solvent or caustic stripping formulations. Perhaps the best method of controlling image defects of this type is to eliminate them as much as possible before the part reaches touch-up. In the author's experience, over 90 percent of these defects are due to contamination of the phototool, the printing frame, or the Mylar resist covering during exposure. If these sources are controlled, touch-up of this type will be all but eliminated, provided proper handling to prevent actual physical damage is used.

The other type of touch-up, the removal of extraneous resist, can do more harm than good if care is not exercised. Most dry-film images are thick and brittle. When removing extraneous bits of resist from the nonimage areas with a sharp instrument, it is easy to damage the image area. Proper instruction to touch-up personnel is essential. Touch-up of this type is often traceable to bits of exposed resist in the processor solvent redepositing and adhering to the nonimage metal areas, bridging fine spacings, or adhering inside holes. Another source is scratches in or damage to the opaque areas of the phototool. Touch-up of this type can be minimized by phototool inspection, maintaining clean filters in the processors, and using an opaque frame around the trim areas of the parts. The last prevents small bits of exposed resist from being formed. It is readily incorporated into the photomaster layout.

In summary, four things can be done to reduce or eliminate touch-up.

1. Inspect and clean printing frames regularly to remove bits of resist and other contamination.
2. Inspect phototools periodically for contamination in the clear areas, especially bits of resist, and damage which renders any portions of the opaque areas transparent. Clean and repair the phototool or replace it as necessary.
3. Use reasonable care in handling parts after coating and through exposure to avoid contamination of the Mylar surface. After developing, handle carefully to avoid damage to the image.
4. Maintain a clean processor and filter.

The frequency of the inspections should be based on the quality of the parts. Some prefer not to touch up inexpensive parts, but rather to scan them and judge whether processing defects in the image will cause the part to be lost in subsequent processing, e.g., etching, or whether to strip and reapply the coating.

Baking

Baking is usually unnecessary. Preexposure baking may actually be harmful by vaporizing volatile resist constituents and entrapping them between the Mylar overlay and the part surface. Preheating the part before lamination is preferable and gives better results. Postdevelopment baking of the image is useful in improving adhesion and chemical resistance somewhat if the cycle is not excessive. Good results can generally be obtained at 180°F for 10 to 30 min. If the temperature is too high, adhesion may be reduced rather than improved. Baking immediately after exposure is discussed under the "Exposure" section in this chapter.

When postdevelopment baking is used, it is advisable to first touch up the part so as to harden touch-up inks as well as the photoresist.

Multiple Coatings

Multiple coatings are useful in some electroforming applications that require thick plating. For thickness up to 5 mils, one coating can be used (Riston 50). Above that, multiple coatings are required. Three coatings or less can be processed without too much difficulty. Coating is relatively simple. The initial coat is applied normally. A second coat can be applied directly on the first by simply removing the Mylar and laminating normally. It is advisable to expose at this point using about $1\frac{1}{2}$ normal time for the resist. A short (about 10 min) bake at 180°F followed by a cooling period (about 30 min) with the Mylar surface in place is desirable before laminating on the third coating, if one is necessary, or developing. After applying the third coating, the phototool must be realigned and normal exposure made. Needless to say, alignment must be accurate. After the final exposure, a repeat of the bake-and-hold cycle will render the part ready for developing.

Developing will usually require two or more passes. The first pass should not be made with the water rinse on. The time used initially should be about twice the normal time. Inspection and judgment will dictate the final pass setting, but the water rinse should be used. If the part has an image on one side only, it should be developed face down to reduce pooling of the developer and rinse water. Immediately prior to use, the part should be electrocleaned to remove any scum or bleed-out.

Liquefied Forms of Dry-Film Resists (Primers)

The unusual application which employs liquefied forms of dry-film resists may be useful on scratched, rough, or uneven surfaces; on surfaces where adhesion has otherwise proved difficult; and in instances where the additional time expended is justified by the value of the part. In essence, a very thin layer of the resist is applied in liquid form over the surface of the part as an adherent base for the laminated resist. As such, the wet film need not be uniform or continuous, may have a substantial number of pinholes and defects, but must not contain residual solvents. The primed surface, in fact, should be as dry as possible.

First, it is necessary to prepare a varnish from the dry film and a suitable solvent. The method of application may have some influence on the solvents selected. Spraying is perhaps the best application method, but satisfactory results have been obtained with a natural-bristle paint brush. One solvent system that works on most dry films is an equal volume mix of xylene and dioxane. Depending upon the resist, several square feet can be dissolved in 100 ml of this mixture. For spraying, xylene can be used to thin the resist as required. The resist used to formulate the primer is generally the one which will be laminated over the surface. However, Riston 30S has been used to prime surfaces for a variety of Du Pont resists, as has Laminar HS 2.0 mils thick. If we use Riston 30S as an example, about 3 ft^2 (24 to 25 g) is dissolved in 100 cm^3 of solvent. For spraying, a solution of 5 to 20 ml of concentrate diluted to 100 ml with xylene will give adequate results. The concentrate can be sealed in dark glass bottles for several months with no apparent ill effects.

After coating, the primer should be baked 5 to 10 min at 80 to 105°C to dry. At this point, the dry film may be laminated over the coating, or a secondary image can be formed by exposing and developing the primer before lamination. When exposing, which is the preferable choice, an exposure time about equal to the full thickness of dry-film resists is required despite the thinness of the film. Oxygen effects cause a substantial loss of reciprocity during exposure. If the dry film is exposed together with the primer, the exposure will be longer than normal by perhaps 20 percent or more. Developing is generally longer also. It is advisable to run a pass at the normal time, but without the water rinse, and then inspect the part. Thereafter, a second pass with the rinse on can be used to clean up any residuals.

PHOTORESIST REMOVAL

After the resist has served its purpose, it must be removed before subsequent processing. Considering the importance of the operation, stripping has received scant attention. Many problems can result downstream with poor stripping that are not readily traceable back to the operation. For example, lamination of improperly stripped multilayer details containing residual dry-film resists into a final board will usually have a good appearance, even if there is enough resist left to give a coloration above the residual-resist-coated areas. However, upon hot oil flowing of the solder plating or similar high-temperature exposure, fine white-appearing blisters will appear in the laminate from the photoresist.

There are many types of photoresist strippers commercially available. Table 6-10 shows the three basic types as classified by the author. Not included are the solvents and azeotropic solvent-stripping units, such as the one shown in Figs. 6-21 and 6-20. The former contains either methylene chloride alone or, in some instances, an azeotrope with a lower alcohol. Also not included are multiple-step stripping operations. Combining different strippers listed in the table could be useful if base-material compatibility could be maintained.

TABLE 6-10 Type of Photoresist Strippers

Type	Stripping action	Composition base	Uses
TYPE I. Room-temperature solvent strippers compatible with plastic board material	After a short soak, it penetrates, swells film, and causes loss of resist adhesion. Some dissolution of the resist also occurs. Mechanical scrubbing nearly always required.	Methylene chloride base solvent with performance enhancers	Should be compatible and usable for all types of resists and finishes if properly formulated.
TYPE II. Heatable solvent strippers compatible with plastic board material	Same as Type I, but generally more effective because of the higher temperature. Mechanical scrubbing may or may not be required. Operated generally from room temperature to 130°F.	Varies. May contain chlorinated benzenes, glycol ethers, glycol ether acetates, chlorinated aliphatic hydrocarbons, formic and acetic acids, and other higher-boiling solvents.	Generally formulated for specific types of resists. Can be used with exposed epoxy or other board material or metal resists.
TYPE III. Hot-alkaline or acid strippers, incompatible with plastic board material	Soaking will chemically destroy and dissolve most of the resist films through hydrolysis or other form of chemical action. Generally used hot (>130°F), with no mechanical scrubbing required.	Typically contains chlorinated phenols or cresols in concentrated aqueous caustic solutions with glycol ethers and sometimes deesterifying additives.	Used in chemical milling and for stripping resists from pattern-plated boards with no exposed epoxy or other plastic.

Type I Strippers

By far the most widely used stripper is the Type I. It is the oldest and most versatile for use in small shops where equipment investments must be held to a minimum. Many are based on early paint removers (29–31). Immersion and hand scrubbing are usually employed with these strippers. Simple as the operation may be, there are many considerations to the use of these strippers. One of the primary problems is their volatility and rapid cooling upon evaporation of the solvents. This interferes with the stripping action and results in an imbalance of components. Mostly, the active ingredient (methylene chloride) is lost. Some manufacturers add a small amount of wax (about 0.1 percent), which gathers near the surface to retard evaporation. However, without proper subsequent processing the wax may cause lamination problems if the part is a multilayer detail. There does not appear to be a real answer for this problem.

Using low-surface-area high-volume tanks is helpful. Higher-molecular-weight aliphatic ketones may function to both enhance stripping action and retard evaporation.

Before selecting and using a Type I stripper, it is advisable to evaluate several formulations using specific criteria relating to the intended use. Salient points to evaluate include (1) stripping ability, (2) material compatibility, (3) solution stability, and (4) physical properties.

Another consideration would relate to specific needs at the facility, such as cost, equipment limitations, the need to strip resists other than dry films, and availability. These will not be discussed and are left to the reader to assess, for obvious reasons. Following will be a brief description of possible evaluation procedures. Type I strippers remove all types of resists.

Stripping Ability. With a 30-s soak, the Type I stripper should remove any dry-film resist in less than an additional minute with light brushing. In other words, no more than 1 min 30 s should be required to strip any dry-film resist. Actually the entire stripping should be completed in less than 1 min with fresh stripper. The capacity of the stripper should also be determined if possible. Capacity will be time-related because of the aforementioned evaporation problem, which will drop both the temperatures of the stripper and reduce the volume of the active ingredient. Therefore, to estimate the capacity of a stripper per unit volume (e.g., per gallon), the parts should be run through at a constant rate. One part every 3 min until the parts will not strip thoroughly in 1 min 30 s is an example. The parts should be the same as should be the actual facility when one does comparative evaluations. Useful but less essential tests include evaporation-rate determination and the rate of temperature decrease to equilibrium under static conditions.

Material Compatibility. While it is important to remove the resist rapidly and thoroughly, the stripper still could not be used if it severely damages the part in the removal process. In actual practice, the stripper should have a generous safety factor with regard to material compatibility. Or, stated another way, the part should not be damaged by exposure to the stripper for a time much greater than required to remove the resist. There will be two basic kinds of finishes that the stripper will see: metallic, such as solder when the part is pattern-plated, and organic, such as the exposed board material when the part is photoetched. Obviously, the plastic board material is more susceptible to damage by the Type I stripper than metallic finishes, but some strippers are acid enough to cause solder smutting and damage the finish sufficiently to promote etching defects.

The test itself is quite simple and quick. Specimens should be prepared by taking samples of copper-clad board material and etching away the cladding. Do not use unclad material because the surface finish is not representative. The size of the part is unimportant, but it should be large enough to protrude from the stripper container. Because the stripper should remove the photoresist in 1.5 min, there should be no significant visible attack in a 3-min static soak (specimens should be thoroughly water-rinsed and dried before examining). The extent of the attack should be determined by comparing the end of the specimen immersed in the stripper with the other end. If it is judged that stripper damage is not significant, a second part should be immersed for 6 min and examined. A final test for 10 min may be useful if several formulations show no significant damage at 6 min. In the author's view and experience, testing Type I strippers past 10 min is of little value. The primary reason for the suggested 100 percent safety factor is to accommodate variations in board material, which will occur. As a general rule, parts may be restripped one time in strippers which do not damage specimens significantly in 6 min.

Solution Stability. One characteristic of Type I strippers that is important is the tendency to separate upon setting. Strippers are made up of general components including a wetting agent and low-specific-gravity components, such as alcohols and ketones. The bulk component is methylene chloride, which is a high-specific-gravity compound. Upon setting, layers will form, and the wetting agent will float on the top in beads, usually dark brown. To be effective, the components must be mixed thoroughly. This is sometimes difficult because of the problem in mixing large drums. It is possible to formulate at a reasonable cost properly coupled solutions that do not readily separate upon storage. A simple test can be performed to determine the stability of the stripper mix. To start with, get a 1-gal sample from the suppliers, or any size that can be thoroughly mixed. Getting a sample from a stagnant drum has the obvious disadvantage of being an already separated sample. After thoroughly mixing, pour a sample into a clear, clean glass container, and seal it tightly with a compatible cap to prevent evaporation. After it sits overnight, look for layers, especially the wetting agent. However, poor component coupling can sometimes be overcome by procuring the stripper in small containers which can readily be mixed before pouring.

Physical Properties. Some Type I formulations may have undesirable physical properties. For example, a component such as formic acid will be unusually irritating to the eyes, nose, and skin. Other examples of undesirable physical properties include nauseating or disgusting odors and a tendency to be absorbed and retained by the board material. Phenols are particularly difficult to remove completely from epoxy surfaces without baking and may cause multilayer board lamination problems if left in place. The stripper should also rinse freely in water.

One important property is the tendency of the stripper to burn or explode.

A flash-point determination is therefore highly recommended. Specific gravity, color, acidity, and the like may be useful as procurement controls along with the material compatibility and stability tests.

A Basic Hand Stripping Process for Type I Strippers. To summarize, the following procedure is suggested for hand stripping with Type I strippers:

1. Soak the part in the stripper by total immersion for 30 s.
2. Remove and brush lightly with a soft-bristle brush to remove the resist. Keep the part wet with stripper from the soak tank.
3. Rinse thoroughly with fresh stripper in a second immersion tank for about 15 s.
4. Water-rinse thoroughly. A spray is preferable.
5. Air dry and bake (200 to 250°F for 15 to 30 min).

The area should be well ventilated and chemically resistant gloves worn by all personnel.

Type II Strippers

There is too great a variation in the stripper formulations to suggest a process, as with the Type I stripper. Some rinse well and may even be water-soluble, such as the glycol ether bases. Others, such as those containing chlorinated benzenes or phenols, rinse very poorly and may form milky precipitates in the rinse water and on the parts. With the latter types, it is usually advisable to follow their immersion with a short dip (<30 s) in a Type I stripper that is particularly free rinsing. Generally, it is always advisable to follow the Type II stripping operation with a bake to expel the higher-boiling solvents.

Evaluation of these strippers is similar to that outlined for the Type I strippers. An important exception is the times given. Often Type II strippers will require longer than $1\frac{1}{2}$ min to strip the resist. Strippers should simply not attack the board material or metallic finish in twice the stripping time.

Examples of Type II strippers are NA19, manufactured by J and S Labs of Seal Beach, California; Leeder 461-W-1, manufactured by Leeder Chemical Company, Paramount, California; and Du Pont stripper 1000.

One general advantage of Type II strippers is their high capacity for stripping dry-film resists. Stripping is usually accomplished by placing the parts in racks and soaking them for several minutes (up to 10 min). The softened resist usually sloughs off in the solution, and, depending on how well it rinses, a coupling Type I stripper or a direct water spray rinse may follow. As a safe rule, parts processed through Type II strippers should be baked after stripping at 200 to 250°F to remove residual solvents.

Type III Strippers

Type III strippers should be handled very carefully. At operating temperature, contact with the skin will often cause severe burns because of the temperature and chemical action. Parts are racked and soaked in these strippers for a time sufficient to remove the photoresist. Following the stripping, the rack is transferred to a hot-water soak followed by a flowing soak or spray. The hot rinse is essential to the operation. Often, the Type III strippers are supersaturated at room temperature or below and will precipitate on the part in cold rinse water. This effect is greatly reduced with the hot-water rinse. For best results, the hot water should be slightly hotter than the stripper and should be changed frequently or be flowing at a low rate.

Two characteristics that make the Type III stripper different from the Type II are its dissolving stripping action and its incompatibility with organic materials, such as epoxy. This relegates it to stripping plated boards or solid metal parts. It exists in both acid and alkaline forms, but generally the latter is preferred. Some chemical milling shops use concentrated sulfuric acid or sulfuric-phosphoric acid mixtures at various temperatures, which would surely qualify as Type III strippers. They are dangerous to use and incompatible with wet parts. An example of a Type III stripper is HAS-27, manufactured by Dynachem Corporation of Santa Fe Springs, California.

Conveyerized Strippers

Conveyerized strippers equipped with stills spray hot solvent on the parts. The resist is swelled and eroded away by the action of the spray. Methylene chloride is an effective solvent, but heretofore its extreme volatility precluded heating. In fact, it is difficult to maintain the solvent at even ambient temperature. Its volatility is so great that it extracts heat from the surroundings as it evaporates and cools below the ambient temperature. In the closed loop stripping machines (Fig. 6-21), it can successfully be used warm. The use of a still (Fig. 6-20) makes stripping inexpensive also. Per square foot, it is probably the most inexpensive stripping method because of the low material and labor cost.

There is a tendency in some instances for strippers of this type to leave a film of adhesion promoter incompletely removed or to deposit bits of undissolved resist in the plated through holes. Because of the expense, it is advisable to run tests to determine the suitability of the equipment for individual needs.

Aqueous-developing Resists

The stripping information discussed thus far is applicable to both solvent- and aqueous-developing resists. The latter resists may also be stripped in

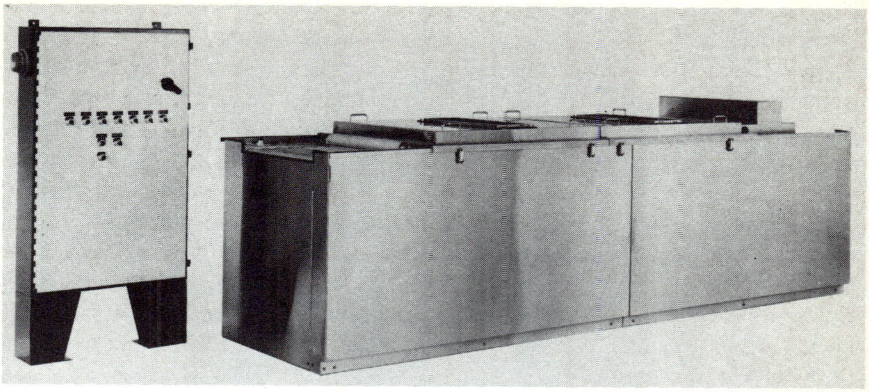

Fig. 6-21 Conveyerized solvent spray stripper. (*Riston Products Div. of Du Pont, Wilmington, Del.*)

equipment and strippers designed especially for them. Laminar A can be removed in strong alkaline cleaners. Laminar AX and Riston II are best stripped by the Type I or II strippers. Figure 6-22 shows the Du Pont ST-20 stripper tank designed for use with the Stripper 1000. Although Stripper 1000 was mentioned as a Type II stripper because it removes the solvent-developing resists, it has compatibility with epoxy and most other circuit-board material and low volatility. It is especially effective for use with Riston II or Laminar AX.

Stripping Criteria

The question arises as to how clean the stripped part must be. It would be desirable to have the part absolutely clean of all residues, discolorations, and stains. However, the degree of cleanliness actually depends upon the subsequent cleaning operations and part disposition Often, light powdery residues from the adhesion promoters of some resists, which can be removed by brushing lightly, will remain on parts. Water stains or oxides from baking may also be present on the metal surface or part. Everything considered, there are three dispositions for the stripped part. These are (1) etching, (2) lamination into a multilayer board, and (3) plating. In the author's experience, the list is in order of the cleanliness required. If the part is to be etched, it will normally see no more cleaning operations. Moreover, deposits that are penetrated by the etchant will etch more slowly than clean areas, and an uneven etch can result. Etchants vary in their sensitivity to penetrate scum. Some are more effective than others. Perhaps the best is the chromic-sulfuric acid and the worst ammonium persulfate. Parts with fine lines and spaces on thick copper must be very clean to etch for best results, whereas the wide lines and spaces can be successfully etched. Wide lines and thin copper are useful to

Fig. 6-22 Spray stripper for use with racked parts. (*Riston Products Div. of Du Pont, Wilmington, Del.*)

accommodate the overetching that may be required to clean up a scummed part. Fine spacings have a greater tendency to electrically short than wide spaces, which is a serious problem in multilayer-board manufacturer.

If the stripped part is a detail or intermediate layer that will be laminated into a final multilayer board, it will undergo some form of subsequent cleaning. This may vary from no more than a solvent wipe or pumice scrub to an elaborate copper conditioning operation, but in nearly all instances it is adequate to remove light residues from metal surfaces.

It is somewhat rare for a process to require plating after removal of photoresist. If it should occur, cleaning processes in a good plating line are generally adequate to remove light stripping residues.

To summarize, the stripping operation must economically serve the needs of subsequent operations. It is an added expense to get parts unnecessarily clean if the subsequent processing is tolerant to loosely adhering residue, but the stripping operation should never leave any adherent organic films. To reduce stripping problems, there are some precautions which can be useful. For one thing, strip the parts as soon as practical after processing. Prolonged storage (several days), especially after long plating cycles, seems to enhance the adhesion of the resist and make stripping more difficult. Postbaking should be avoided with high-adhesion resists. Use the thinnest resist practical. Not only does this reduce resist costs but stripping costs as well. As a general rule, anything that promotes polymerization of the resist works against stripping.

OTHER MATERIALS

A variety of photosensitive resins are available from W. R. Grace and Company. Dry-film photosensitive solder maskant has been field-tested successfully by Grace. These photopolymers utilize sulfide curing elements that cure by free-radical reactions. A ketone sensitizer reacts with a sulfide photoinitiator to generate the free radicals. Dynachem also offers a liquid resist that is 100 percent solids that cures from a liquid to a solid upon exposure (the DCR LS series of resists).

OXYGEN EFFECTS—THEORETICAL CONSIDERATIONS

Mathematical studies of oxygen consumption in dry-film photopolymers have been made by Chong (32) and Smith (33) of Du Pont. The depth of the studies is beyond the scope of this book, but for those with mathematical interest the studies will be of interest.

REFERENCES

1. L. R. Plambeck, Jr., U.S. Patent 2,760,863 (1956).
2. L. R. Plambeck, Jr., U.S. Patent 2,791,504 (1957).
3. J. R. Celeste, U.S. Patent 3,469,982 (1969).
4. A. C. Schoenthaler, U.S. Patent 3,418,295 (1968).
5. J. F. Herzog, *Proc. EIPC Conf. Printed Circuit Manufacturing*, Feb. 19–20, 1970, pp. 108–115.
6. J. R. Celeste and S. Bauer, U.S. Patent 3,261,686 (1966). Also see Cohen et al., U.S. Patent 3,380,831 (1961).
7. N. T. Notley, U.S. Patent 2,951,758 (1960).
8. G. Oster, U.S. Patent 2,850,445 (1958).
9. G. Oster, U.S. Patent 2,875,047 (1959).
10. G. Oster, U.S. Patent 3,097,096 (1963).
11. G. K. Oster and G. Oster, U.S. Patent 3,074,794 (1963).
12. G. K. Oster and G. Oster, U.S. Patent 3,097,097 (1963).
13. G. K. Oster and G. Oster, U.S. Patent 3,145,104 (1964).
14. R. B. Heiart, U.S. Patent 3,060,026 (1962).
15. F. P. Alles, U.S. Patent 3,458,311 (1969).
16. J. R. Celeste, U.S. Patent 3,448,089 (1969).
17. F. P. Alles, U.S. Patent 3,475,171 (1969).
18. M. Burg, U.S. Patents 3,060,023 and 3,060,025 (1962).

19. F. P. Alles and W. R. Sauer, U.S. Patent 2,627,088 (1953).
20. Du Pont, British Patent 905,700 (1962).
21. L. J. Miller, J. D. Margerum, and J. B. Rust, *J. SMPTE*, vol. 77, pp. 1177–1184, November, 1968.
22. P. Walker, V. J. Webers, and G. A. Thommes, *J. Photographic Science*, vol. 18, pp. 150–158, 1970.
23. D. W. Woodward, V. C. Chambers, and A. B. Cohen, *Photographic Science and Eng.*, vol. 7, no. 6, November–December, 1963.
24. Author's data.
25. H. J. Bachtel, U.S. Patent 2,811,252 (1957).
26. D. E. Rapp, U.S. Patent 2,923,747 (1960).
27. W. E. Brown, U.S. Patent 3,049,571 (1962).
28. J. Lockhart, *Minutes of California Circuits Association 1971 Fall Symposium*, pp. 32–45.
29. Federal Specification TT-R-00251.
30. Federal Specification TT-248.
31. Federal Specification TT-243.
32. J. S. Chong, *J. Applied Polymer Science*, vol. 13, pp. 241–247, 1969.
33. D. K. Smith, *Photographic Science and Eng.*, vol. 12, no. 5, September–October, 1968.
34. J. R. Celeste, U.S. Patent 3,526,504 (1970).
35. C. T. Chang, U.S. Patent 3,549,367 (1970).
36. V. C. Chambers, U.S. Patent 3,479,185 (1969).
37. R. B. Hurley and S. Lazarus, U.S. Patent 3,622,334 (1971).
38. L. Roos, U.S. Patent 3,758,306 (1973).

CHAPTER SEVEN

Microelectronic Applications

Photoresist in microelectronics is a viable subject for a book in itself. It is there that the photoresist art and science reach the ultimate, and the treatment herein will necessarily be brief.

There is a mild controversy regarding the merits of positive resists versus the negative types in microelectronics, and there are "positive shops" and "negative shops." Many use both types of resists. Each has its advocates and its adversaries, and direct comparisons are quite difficult. Those seeking to find a preference in this chapter will be disappointed, but both types will be considered from a processing standpoint.

As in the other chapters, a basic process is given (Table 7-1) and thereafter related subject matter is discussed. Initially, a brief review of the

TABLE 7-1 Typical Process Sequence

1. Prepare substrate	5. Develop
2. Coat substrate	6. Rinse
3. Prebake	7. Dry
4. Expose	8. Postbake

positive- and negative-resist reactions will be given followed by processes.

Microelectronic applications, as used herein, refer to silicon or germanium devices, thin films on ceramic substrates, and other electronic applications where line widths and spacings are frequently below 2 mils (0.002 in.).

NEGATIVE RESISTS

The most widely used resist resins for microelectronic resist applications appear to be the polyisoprene derivatives. To be useful as a resist, *cis*-polyisoprene undergoes a process called "cyclization" in which the level of unsaturation is reduced. This converts the rubber to a hard, resinous material (see Chap. 2). Investigations by IBM (1) and Hunt (2) indicate that an unsaturation level of approximately 6 mmol/g will give a high level of adhesion to silica surfaces. Cyclized rubber is intrinsically photosensitive, a fact known for many years. Because of its excellent dielectric properties, it was often used as an electrical insulator, but it became embrittled from polymerization by sunlight. Rubber cyclization has been known since the 1920s (3), but it was patented as a photoresist resin in combination with azide sensitizers by Kodak (4) in the 1950s. Characteristically, resists of this type are highly resistant to both acid and alkaline aqueous solutions and oxidizing agents, generally more so than the corresponding thickness of a positive-resist film.

The KPR series of resists (including KPL) have also been mentioned for use in microelectronics (5–7). Processing techniques are not so well publicized for these resists as for the isoprenoid resists, and Martinson (8) indicated that greater chemical resistance could be expected from the resin in KMER than from the KPR type of resin. Kelly (5) reported, however, that KPR-2 gives a lower incidence of pinholing than KTFR, whereas a 4 : 1 volume mixture of KPR-KPL gives results similar to KTFR at approximately the same thickness.

Information regarding photoinitiators or sensitizers is, of course, limited. Kornfeld (9) indicated that the KMER sensitizer is an aromatic *bis*-azide (Fig. 7-1). The sensitizer would also work with KTFR, which has a similar resin composition. Compounds of this type are known to undergo reactions which involve the removal of certain hydrogen atoms from the resin chain, with the formation of a chemical bond to the chain through the nitrogen (Fig. 7-2). Free radicals are generated by the sensitizer upon exposure, which can react in a number of ways besides the one shown. The unsaturated linkage in the sensitizer may also be polymerizable by heat for further cross linking. KPR-2 was analyzed and shown to have a sulfur-bearing sensitizer (10). Later patent literature (11) indicates that the sensitizers from the Robertson and West patent (12) may be employed.

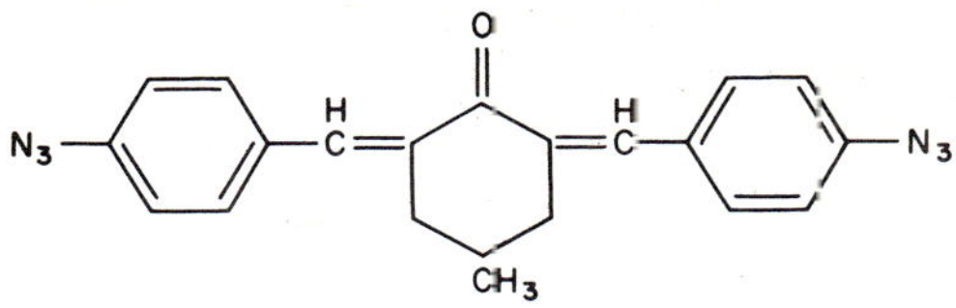

2,6 Di-(4'-azidobenzal)-4 methyl cyclohexanone
(Sagusa and van Allen, U.S. Patent 2,940,853, 1960)

Fig. 7-1 Sensitizer used to polymerize cyclized-rubber resists (aromatic azide).

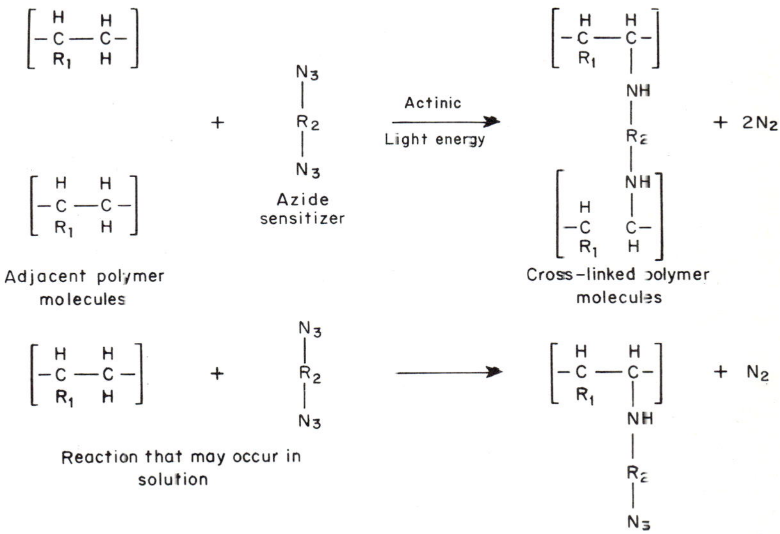

Fig. 7-2 Some reactions of azide sensitizers.

POSITIVE RESISTS

Although the stepwise processing sequence is similar to negative resists, the chemical concepts of positive resists are drastically different. Exposure to light hardens negative resists by a propagating mechanism which greatly increases the molecular weight of the exposed areas. Despite the increase in molecular weight, the developing solvents of negative resists which are capable of dissolving the unexposed areas will also be readily absorbed by the exposed areas, and swelling will result. With swelling comes some image distortion, which is unavoidable. Positive resists, however, undergo no polymerization reaction, and the developing solution is not normally absorbed by the image areas. Image distortion is thereby minimized during developing. The actual reaction which results in image formation involves the conversion of functional groups in the sensitizer from a hydrophobic diazide to a group which is soluble in dilute aqueous alkaline solutions through neutralization and salt formation. Water does not tend to swell or distort the image and is an effective solvent for the neutralized carboxylic acid groups after developing.

During exposure, one of the reactions shown in Chap. 2 occurs. There are no chain-propagating reactions from free radicals, such as those formed during the exposure of negative resists. Radiation does not tend to cause the formation of a substance capable of reacting outside the sensitizer molecule to which it is attached. There is also no apparent reaction with the resin by the sensitizer during exposure. Oxygen does not significantly interfere with this reaction, as it does with the free-radical type of reaction that occurs during the exposure of negative resists.

Since image-distorting phenomena are minimized with positive resists, it is possible to achieve exceptional resolution and edge acuity with positive resists in comparatively thick coatings. It has been reported (13) that the image deviation from the photomask is independent of coating thickness for some of the Shipley resists, but the study did not involve the tolerances encountered in microelectronics. Nevertheless, positive resists can produce very fine geometrics in relatively thick coating with a minimum of technique.

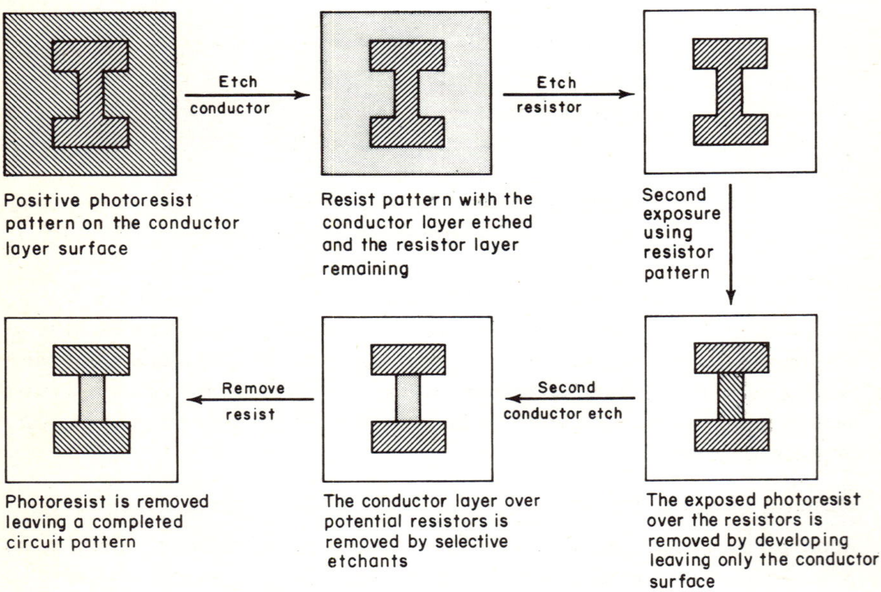

Fig. 7-3 Example of the double-exposure process with a positive resist.

After image formation, the positive-photoresist image retains light sensitivity. Using this property as an advantage, it is possible to save an entire photoresist operation in some instances. Figure 7-3 shows an example of how a second exposure was used in photoetching thin films to avoid an entire photoresist operation (14).

Positive resists do not possess the inherent chemical resistance of the negative resists, particularly to alkaline and oxidizing solutions, and are more expensive to procure. Some operations use positive resists where possible and resort to the negative resists for the jobs requiring greater chemical resistance, although high-temperature postbaking (150 to 200°C) greatly increases the chemical resistance of positive resists.

SUBSTRATE CLEANING

The photoresist layer must be physically close to the substrate surface for a good adhesion. The secondary attractive forces that are believed to be responsible for adhesion are effective only over very minute distances. The presence of contaminants such as moisture, incompatible organic substances such as oil, or loosely adhering inorganic residues can prevent the intimate surface contact required for good adhesion. Moreover, contamination causes poor surface wetting of the resist by reducing or neutralizing the attraction the resist molecules have for the surface. To be suitable, a cleaning process must be capable of removing organic and inorganic contaminants, moisture, and particulate matter and leave a surface receptive to the photoresist. Needless to say, it must also be compatible with the substrate material.

Cleaning operations vary in their complexity, and many are proprietary or even patented in some instances. Generally, however, the overall operation contains three processes in sequence:

1. *General cleaning processes.* This operation removes organic soils, inorganic contaminants, and particulate matter already on the part.
2. *Dehydration bake.* Moisture and volatile cleaning residues are removed from the surface to which the resist will be applied.
3. *Final or whirler cleaning process.* Here, any small residuals from previous cleaning operations are removed just prior to resist application.

This sequence with its variations can be used to prepare nearly any substrate for resist application. Following will be discussions and examples of each process.

General Cleaning Processes

General cleaning processes actually constitute the bulk of the cleaning that will be performed to remove contaminants from previous operations. Therefore, they should be based to some extent on the removal of residues that may have been left from earlier processing, such as ash, oil films, residual photoresist, etc. However, one capability that should always be present in these operations is degreasing.

Cleaning operations must not replace one form of contamination with another. The methods and chemicals employed must be selected with this in mind. Several types of devices are affected by sodium contamination, and many ordinary chemicals contain sodium even in a highly purified form. Table 7-2 gives some examples of the sodium levels in some pro-

TABLE 7-2 Sodium Contents Noted on Some Processing Chemicals (15)

Sodium content	Chemical
High (>1,000 ppb)	CrO_3, H_3PO_4, AZ-111 and AZ-1350H photoresists and normal developers
Medium (250–1,000 ppb)	HNO_3, H_2SO_4, KTFR
Low (50–250 ppb)	HF, NH_4F, isopropyl alcohol, acetic acid, acetone
Very low (<50 ppb)	Trichloroethylene, KMER developer and thinner

cessing chemicals as noted by Brandewie and Scott (15). Water is often used in these cleaning processes and should be of high purity. Table 7-3 gives some properties as listed by Monzeglio (16). In addition to the properties listed, microorganism levels, turbidity, and dissolved silica may be necessary criteria. For most applications in microelectronics, water properties approaching the high values are desirable, e.g., $14 > M\Omega$.

Shortly after exposure to air, high-purity deionized water will undergo a drastic change in resistivity. Carbon dioxide is absorbed from the air to

TABLE 7-3 Some Properties of Processing Water for Electronic Applications (16)

Quality level	General	Medium	High
Resistivity ($\Omega \cdot$ cm)	10^5	10^6	$16–18 \times 10^6$
Dissolved solids—max.	4.0 ppm	0.4 pp,	25 ppb
pH range	5.5–7.5	6.5–7.2	6.9–7.1
Total carbon content (ppm) —max.	5.0	2.0	0.1
Total silica content (ppm) —max.	3.0	1.0	0.1
Nonvolatile residues (ppm)(0.22-μm filter)—max.	20.0	10.0	0.5
Total iron content (ppb) —max.	100	10	2

form carbonic acid, which is quite weak as an acid. Still, it ionizes sufficiently to give a resistivity of about 1 MΩ · cm at equilibrium:

$$\text{Absorption: } CO_2(\text{gas}) + H_2O(\text{liquid}) \longrightarrow H_2CO_3$$

$$\text{Ionization: } H_2CO_3(\text{dissolved}) \longrightarrow H^+ + HCO_3^-$$

The rate of change will vary with the surface area–to-volume ratio of the water, the carbon dioxide concentration of the air, temperature, agitation rate, and other factors which would affect gas absorption into a liquid.

There are numerous general cleaning processes which do not utilize water. Table 7-4 gives an example in which warm organic solvents are used. Tables 7-5 and 7-6 give examples of aqueous cleaning processes using powerful oxidants to degrease and dissolve metallic contaminants. The alkaline-peroxide process (Table 7-5) must be started with a fresh

TABLE 7-4 General Cleaning Process with Solvents

1. Soak clean in warm (>60°C) trichloroethylene 2 to 5 min.
2. Rinse in dry isopropyl alcohol for 1 to 2 min at ambient temperature.
3. Dry with nitrogen.

TABLE 7-5 General Cleaning Process with Aqueous Alkaline Solution

1. Place parts in an aqueous 5 vol percent hydrogen peroxide (30% solution) and 10 vol percent ammonium hydroxide.
2. Heat to boiling and boil 30 min.
3. Rinse in overflowing deionized water 5 min.
4. Rinse in isopropyl alcohol 1 min.
5. Dry with nitrogen gas.

TABLE 7-6 General Cleaning Process with Aqueous Acid Solution

1. Soak clean in hot (90 to 120°C), concentrated (68 to 71%) nitric acid 3 to 5 min. Enter *dry* because water or solvents may react violently under these conditions. Drain 5 to 10 s.*
2. Immersion-rinse 2 to 5 min in flowing water.
3. Immersion-rinse 3 to 5 min in flowing water.
4. Soak in 35 to 50 vol percent isopropyl alcohol–water mixture for 30 s. Drain 3 to 5 s.
5. Soak in isopropyl alcohol for 30 s.
6. Dry with nitrogen.

* Many types of substrates, metals, and plastics are reactive with hot, concentrated acids. Glass or Teflon racks and containers are probably best suited for this application. Compatibility tests with substrates are recommended.

solution of peroxide for each batch of parts. Peroxide is decomposed during the cleaning process by boiling, and both cleaning compounds (ammonia and nascent oxygen) are gaseous. Solution contamination possibilities are thereby reduced. This and other processes are described in an early ASTM publication (17). A powerful oxidizing acid in a hot, concentrated form is also an effective cleaner if the substrate can withstand such conditions (Table 7-6). Other examples include Caro's acid formed by mixtures of sulfuric acid and hydrogen peroxide. Chromic-sulfuric acid mixtures are also effective but are difficult to dispose of and may leave chromate finishes on some metals. Sometimes, solvent cleaning is followed by an aqueous detergent cleaning solution, especially when sodium contamination is not a problem. Ultrasonics (Table 7-7) is also used to enhance cleaning action.

TABLE 7-7 General Cleaning Process with Solvents and Ultrasonics

1. Soak clean in warm (60°C) trichloroethylene 2 to 4 min in an ultrasonic unit (40 kHz).
2. Rinse in dry methanol for 1 to 2 min at ambient temperature.
3. Dry with nitrogen gas.

Dehydration Bake

Baking is an essential cleaning operation, especially for silica surfaces. At room temperature, the silica surface has many silanol molecules present which will absorb moisture and retain it past the boiling point (100°C). This is a most undesirable condition, and poor resist adhesion would be at least probable. If the silanol (silicon-alcohol linkage) is heated to 200°C, the absorbed moisture is released, but the silanol remains. Under the previous conditions, a silane coupling agent would be expected to react effectively with a surface dehydration and covalent-bonding reactions (see "Resist Application" section.) At 200°C, there is usually enough dehydration to allow adequate resist adhesion to occur. A more desirable temperature is 600°C, at which the silanol condenses to silica with the evolution of moisture.

Unfortunately, the reactions are reversible. That means simply that hydration will eventually reoccur at reduced temperatures in ambients with moisture present. Since moisture cannot be excluded by practical means, it is necessary that resist be applied soon after heating and that cooling take place in a reasonably dry environment. Applying resist directly to hot surfaces should be avoided because rapid solvent expulsion will cause the resist to prematurely set and uneven coatings can result.

Metal surfaces that hydrate readily, such as aluminum, can also benefit from a dehydrating bake at 200°C. At 400°C in air, many organic substances are decomposed if the duration of the bake time is adequate.

Baking is an especially effective cleaning at temperatures of 400°C or higher and often the only precleaning that may be necessary.

The silanol groups can be present before cleaning on parts previously processed or stored in moisture-containing ambients (18–20).

Final or Whirler Cleaning Cycles

At this final point in the cleaning operation, the primary contaminant is particulate matter or perhaps very thin layers of vaporous materials that may have condensed from the baking operation. Solvents applied to the part should be as nonpolar as possible but still retain compatibility as a dilutant for the resist. Surfaces that have a strong affinity for water, such as aluminum or silicon, should be cleaned with dry solvents. Otherwise, the water may preferentially react with the surface. Only a tiny amount of water in the solvent (0.1 percent) is required. Surfaces such as gold which have little water affinity may be cleaned with solvents of a more hydrophobic nature, but, again, hydrophobic solvents are advisable. Perhaps the most useful all-around solvent for this purpose is xylene, but others are shown in Table 7-8. Nitrogen-gas blowoffs are often used in

TABLE 7-8 Solvents for Cleaning Immediately Prior to Resist Application (Final Cleaning)

Resist type	Example	Solvents for cleaning
Cyclized rubber	Waycoat IC, Kodak Micro resist 747, CMR Series	Thinner, xylene trichloroethylene, toluene, Stoddard solvent, and mixtures thereof
Cinnamate ester resists	KPR-2, KOR	Xylene, Waycoat IC or Kodak Micro resist thinner, cyclohexanone
Positive resists	AZ-1350J, AZ-1350B, PR-102	Xylene, butyl acetate, or AZ thinner, ethyl acetate, xylene, thinner

lieu of or in combination with solvents. The gas must be dry and warm enough to avoid chilling the part and causing atmospheric moisture to condense on its surface. Gas blowoffs are not as effective as liquids in removing particulate matter, but are less expensive and widely used. For effective cleaning, the part should be spinning at high revolutions per minute when the solvent is applied. Waves of fresh solvent will wash over the part with some degree of force to remove particulate matter; this is followed by a spin-dry. This cleaning operation should take place as close as possible to the resist application.

RESIST APPLICATION

Regardless of the type of resist, the ultimate aim is to produce an image of sufficient resolution and chemical resistance to do the job. Several factors must be considered to achieve this aim. Perhaps the most important of these, however, is the resist application process. There is considerable dependence in all subsequent processing on the coating process. The thickness of the coating and thickness variations (uniformity) are controlled by this operation. Prebaking, for example, must be sufficient to remove solvents from a given thickness of resist that could interfere with exposure without extensive thermal polymerization. Thicker coatings require longer times that may be harmful to thin coatings. Exposure is strongly affected by the quantity of resist present. Thicker coatings will require longer exposure times than thinner coatings, if all else is equal, because a greater quantity of resist must be reacted. Poor uniformity will allow varying degrees of exposure on the same part. Chemical resistance and the incidence of pinholing in the image are improved by thicker coatings of the resist. Negative-acting resists depend to a large extent on polymerization for adhesion, which is directly affected by exposure time and the thickness and uniformity indirectly. Developing time varies to a rough extent with the coating thickness, although many resists have a fair tolerance to overdeveloping. Fine lines and high-resolution patterns will require reasonably accurate developing times, however, especially for negative resists, where swelling of the lines may reduce adhesion.

All commercial photoresist formulations have ranges of thicknesses over which a constant set of processing conditions will produce acceptable images. For example, a prebake temperature of 80°C for 20 min may be adequate for fairly thick coatings of resist without causing scum or heat fog on thinner coatings of the same resist. Of course, there are limits, and some steps are more sensitive than others to thickness variations, e.g., exposure. Obviously, many processing problems as well as considerable time and effort can be avoided with resist coatings of repeatable and uniform thickness. Good coatings are the most fundamental starting point for good yields.

In coating-thickness selection, the overriding considerations are the fineness of the lines, particularly the spaces to be resolved, and the chemical resistance required of the image. Generally, negative-acting resists will resolve about 3 times the resist-film thickness without extraordinary techniques, and the positives are somewhat better. Satisfactory chemical resistance of the image has to be determined by testing or, in more obvious situations, the judgment of experienced personnel.

Resist Application by Whirling

There are numerous combinations of spin speeds and resist viscosities that achieve the same average coating thickness. Higher-viscosity mixtures spun at higher speeds can give equal thicknesses to lower-viscosity mixtures spun at lower speeds. For most negative resists, spin speeds from 3,000 to 8,000 rpm or more are favored with viscosities ranging from 20 to over 400 cP (Table 7-9). Positives generally are applied between 3,000 and 6,000 rpm. Resist-film uniformity, the smallest lines and spaces to be resolved, and the chemical (etch) resistance required must be considered when selecting the combination of speed and viscosity, as iterated earlier. Low spinning speeds may lead to edge buildup caused by the surface tension of the resist at the edge of the part, whereas very high speeds may cause thin coatings. In addition to speed, the acceleration, or the time required to achieve a given speed, is also a significant factor in achieving acceptable coating uniformity. Damon (21), in his high-speed photographic study of the effect of whirler acceleration, concluded that high-acceleration whirlers tend to give thinner, more uniform films than low-acceleration types at the same ultimate speed. Thickness differences of as much as 30 percent were reported at the same approximate spinning speed. The thickness difference was attributed to breaking of a surface-tension barrier at the edge of the part early in the process, which allowed more resist to leave. Resist left the center of the part in the form of discrete waves immediately after the surface-tension barrier was broken. Shortly thereafter, wave formation ceased, and a very thin invisible wall formed at the rim of the part; the wall continued to grow in height. Apparently, the high-acceleration whirlers tended to complete the process while the resist was still fluid and avoided edge buildup and poor uniformity.

Dunham (47) indicated that when resist was applied to a stationary part, enough time should be allowed for the resist to displace air from the surface before whirling, or bubbles could be entrapped in the coating. If the part was moving, resist application should continue after the part is visibly coated to ensure air displacement. Some whirlers disperse resist to a stationary wafer and spin slowly to allow time for air displacement until the surface is wetted and then spin rapidly for a preset time.

In Damon's study conducted in 1965, parts (silicon wafers) were flooded with resist before spinning. More recently, there has been a trend toward metering a smaller quantity of resist in the center of the part and spinning.

TABLE 7-9 Examples of Application Parameters

Resist	Use	Resolution (mil)	Viscosity (cP)	Spin parameters: Time (s)	Spin parameters: rpm	Film thickness (Å)
Waycoat IC	General etching	0.2–0.3	43	30	5,000	8500–9500
Waycoat IC	Fine-line etching	0.05–0.1	28	30	5,000	4000–5000
Kodak Micro resist 747	General etching	....	60	30	5,000	8000–8500
KTFR	General etching	0.1	21.5	30	5,000	4900
KMER	MESA etching	....	220	30	1,100	50,000
Waycoat SC	MESA etching	....	450	30	2,000	50,000
			450	30	8,000	25,000
DCR #3170	General etching	0.1	As received	10	4,000	450–650
Four parts KPR, one part KPL	SiO_2 etching	....	....	30	10,000	
AZ–1350J	General etching	....	As received	30	6,000	15,000
AZ–1350B	General etching	....	As received	30	6,000	3000
AZ–111	General etching	....	As received	30	6,000	8000
PR–102	General etching	....	As received	30	3,000	10,000

However, the effect of acceleration on uniformity proved by the study is applicable. Schwartz, (22) also noted that slowly accelerating whirlers produce coatings with poor uniformity.

It is apparent that resist-film thickness should be as great as practical to reduce the incidence of pinholes, but it still must be able to resolve the desired image reliably. Martinson (7) indicated that film thicknesses of 1 μm (10,000 Å) should be adequate for general use down to 0.2-mil lines and spaces using KTFR but that finer lines and spaces could be resolved. Table 7-9 shows some spin parameters for negative resists from various sources that reportedly had sufficient resistance to pinholing for either a specific or general use. Note the tendency toward a 30-s spin time. Clark (23) indicated that a spin speed of 5,000 rpm for 30 s gives uniform coatings and that Shipley AZ-111 or AZ-1350H could resolve 0.1-mil lines and spaces in the resulting film. The former resist gave a film thickness of 7000 Å, and the latter 12,000 Å with resist viscosity as received. Excellent performance was indicated under these conditions.

Before using, resists are usually purified. Filters may not remove large aggregates of high-molecular-weight substances effectively but do remove particulate matter. In some cases large aggregates get through filters by deforming (24) under pressure. Such matter can be removed by centrifuging (24,25), however. Taylor (26,27) indicated that KMER could be purified by electrophoresis using a high dc (several thousand volts) voltage for several days across a solution of KMER and thinner. Perhaps the earliest and most thorough purification was that instigated by Texas Instruments (TI) in the early 1960s for KMER and KTFR. Harrell (25) described the procedure in 1966. Very briefly, the process involves the following operations:

1. Adjust incoming resist viscosity.
2. Chemically treat to precipitate organic impurities.
3. Chemically treat to precipitate inorganic impurities.
4. Centrifuge the resulting liquid 48,000 *g* hours.
5. Adjust final viscosity.

Improved images resulted from the elaborate procedures, and the product was eventually made available to other companies. Before undertaking elaborate purification processes, such as the TI process, it is usually necessary to know the basic resist composition. Otherwise, steps that may selectively remove vital constituents, such as the sensitizer, may result in greater image-property variation than the original unpurified resist.

In recent years, photoresist suppliers have made substantial efforts to remove "junk" from their formulations, and improvements have been noticeable. Newer resists such as the Kodak Micro resists or the Hunt Waycoat IC series normally require no filtration or viscosity control prior to use. Positive resists tend to be pure, also. Little is written about positive-resist purification processes. Generally, filtration and viscosity adjustment prior to application are used.

In addition to resolution, the image must withstand the etchants for the desired length of time. Adhesion loss and pinholing are the two most common problems relating to the resist thickness. Adhesion promoters have been introduced that are applied directly to the substrate surface. Bryant (28) suggested the use of diluted photoresist applied and dried before the application of the regular coating to silicon dioxide surfaces to improve adhesion. A recent Kodak patent (11) indicates that the addition of novolak resin to polyisoprene and polyvinyl cinnamate resists generally improves adhesion and chemical resistance. Under the category of adhesion promoters for silicon dioxide and ceramic surfaces, however, perhaps the significant group of chemicals is organic silicon compounds.

Silane adhesion promoters (29–33) for silica surfaces have long been known. Most of the development has centered on preparing glass cloth laminates, where good resin bonding was essential for strength. The coatings are applied to the glass surface and cured before resin is applied. Recently, interest has been shown in the use of silane coupling agents to improve photoresist adhesion to silica surfaces. These coupling agents contain two types of functional groups attached to a silicon atom. One group is capable of reacting chemically with the substrate, while the other bonds with the coating resin. Under the proper conditions, the substrate-reacting groups react with surface moisture and existing silanol linkages to produce an actual chemical bond. Surface-reacting groups are readily hydrolyzable to yield one or more silanol ($\equiv$SiOH) groups, which can react by condensing with the similar groups on the surface to form a chemical bond. The resulting chemical bond is through an ether linkage ($\equiv$Si—O—Si$\equiv$) with the evolution of water. Examples of the surface-reacting hydrolyzable components of the coupling molecule include alcohol (—OH), chloride (—Cl), and methoxy ($-OCH_3$) groups. Functional groups of this type are believed to be capable of reacting with either the surface silanol or moisture retained by the silanol surface:

Surface moisture present:

$$\mathrm{RSiCl} + \mathrm{H_2O} \longrightarrow \mathrm{RSiOH} + \mathrm{HCl}$$

$$\mathrm{RSiOH} + \mathrm{HO{-}Si(surface)} \longrightarrow \mathrm{RSi{-}O{-}Si(surface)} + \mathrm{H_2O}$$

Silanol surface:

$$\mathrm{RSiCl} + \mathrm{HO{-}Si(surface)} \longrightarrow \mathrm{RSi{-}O{-}Si(surface)} + \mathrm{HCl}$$

Considerable heat is required for the reactions to take place irreversibly. Temperatures near 200°C appear adequate for chlorosilane (33) but very high temperatures in air can cause damage or decomposition, depending upon the nature of the resin-reacting component. Commercial couplers usually contain three of the hydrolyzable groups per molecule to ensure that at least one will be available to react with the surface silanol. The reason for the excess is the probability of interreactions with silanols from other coupling molecules in lieu of the surface silanol. Bartfield (34) reported satisfactory results with a dichlorosilane to improve photoresist adhesion, however.

In addition to compatibility with the resist, the organic functional groups that bond the coating to the surface must not react to produce a surface that is not readily removable by some means, usually the acid fluoride etchant. Bartfield indicated that dimethyldichlorosilane gives satisfactory results and drastically improves resist adhesion. Diphenyldichlorosilane may also be useful. Application solutions are usually quite dilute and are spun on to give a very thin coating. Baking follows. Usually 200°C for 30 min is adequate. The length of time between application and use was not given but generally is as soon as practical.

Collins and Deverse (35) patented an unusual adhesion promoter that could be added directly to the photoresist or precoated on the substrate. The compound cited was hexamethyldisilazane; other hexaalkyldisilazane compounds could also be used. One primary advantage of this type of adhesion promoter is the absence of any high-temperature bake such as that used in the chlorosilanes. In fact, no bake other than the photoresist prebake (100°C for 10 min) was cited. KTFR with 6 percent dibutyl adipate was the resist used in the example. No postbake was used on a silicon wafer, and it was claimed to be useful on both metallic and nonmetallic surfaces with stable oxides. The mechanism was not given, but it was proposed that the disilazane reacts with the surface oxide in some manner.

Some resists possess sufficient tackiness to cause cobwebbing. While not throughly understood, some factors are known. Upon reaching the edge of the substrate, portions of the resists are thrown off by centrifugal force. The force, which is a function of the spin speed of the substrate, causes the liquid to flow without overcoming the adhesive forces. Cohesive forces within the liquid then form fine threads which have the appearance of cobwebs around the spinner cup. Besides the spin speed (revolutions per minute), the solids content of the resist (or the dilution factor), the molecular weight of the polymer, and the viscosity are all factors that relate to this phenomenon. For a given resist, lowering the viscosity or reducing the overall solids content by dilution or thinning would be helpful, as would reducing the spinner revolutions per minute.

Spraying has been used to some extent for microelectronic resist application. Because the parts are static, no residual stresses in the coating are encountered as in spinning, and the output is much greater. A special spray unit is manufactured by Zicon[1] contained within a clean hood which reportedly applies uniform spray coatings of positive or negative resist as thin as 3500 Å, using a vaporized organic liquid as the spray carrier gas.

Miniature roller coaters (Fig. 7-4) with special rollers for microelectronic use have recently been developed by Gyrex Corporation.[2] Like spraying, the output of parts is much greater than with spinning operations.

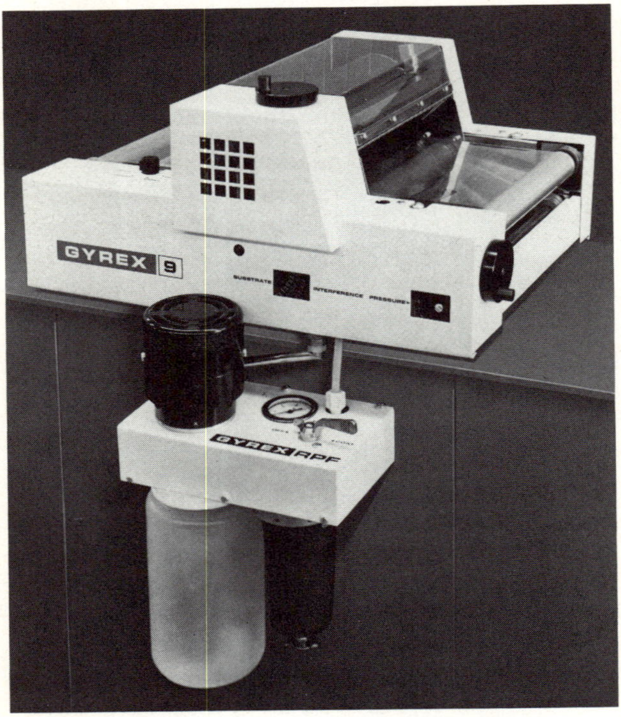

Fig. 7-4 Roller coater for microelectronic applications, the Gyrex 9. (*Gyrex Corp., Santa Barbara, Calif.*)

Though relatively new, some units are in production use, and the units are comparatively inexpensive to procure. One advantage of the roller coater over other methods is the reduced waste of resist. Losses in whirling reach 80 to 90 percent if the substrate is flooded and 50 percent or more with many spray applications. Because the roller coater is conveyerized, infrared drying ovens can be attached immediately thereafter for rapid and complete resist drying (Fig. 7-5). Both positive and negative resists can

[1] Zicon Corporation, Mount Vernon, N.Y.

[2] Gyrex Corporation, Santa Barbara, Calif.

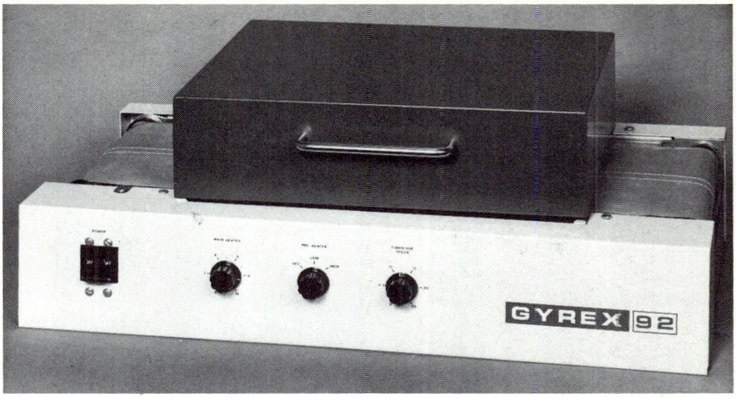

Fig. 7-5 Conveyerizer photoresist drying oven for microelectronic applications. (*Gyrex Corp., Santa Barbara, Calif.*)

be applied including resists not normally roller-coated, such as Shipley AZ-1350H. Stockley (36) gave the values shown in Table 7-10 for roller-coating AZ photoresists. A roller coater similar to the one shown in Fig. 7-4 was used at a conveyer speed of 5 ft/min. Adjustments are critical, however, to maintain thin, uniform coatings. A roller coater is used quite successfully in the manufacture of thin-film hybrid circuits.

TABLE 7-10 Microroller Coating Thickness with Positive Resists (36)

		Coating thickness (μm)			
		Coarse roller		Fine roller	
Resist	Viscosity as received (cP)	Single pass	Double pass	Single pass	Double pass
AZ-111	36	2.0	2.4	0.5	0.9
AZ-1350	4.5	1.5	2.1	0.4	0.5
AZ-1350H	25	2.8	3.1	2.0	2.4

COATING SPEED: 5 ft/min. COARSE ROLLER: 96 threads per inch. FINE ROLLER: 184 threads per inch. NOTE: Additional data are available from Gyrex Corporation, Santa Barbara, Calif.

Prebaking and Drying

Although the actual bake time and temperatures can vary significantly within certain limits, the need for prebaking is critical. The actual purpose of the prebake is threefold:

1. Eliminate solvents from the films to ensure repeatable exposures.

2. Increase the adhesion to the substrate to better endure the developing cycle without excessive damage or distortion to the image area.
3. Harden the film to resist physical damage from handling and processing.

Both positive and negative resists need this operation, which should be preceded by a room-temperature air-drying operation.

Air drying before baking allows the retained solvents to evaporate from the film slowly. This tends to prevent the formation of a thin barrier of dried resist on the surface, which may tend to retard evaporation during baking. Solvent retention of this type tends to be greater with negative resists than with positives. The higher-molecular-weight resins in the negative resists are probably responsible. Normally, 10 to 30 min ambient air drying is adequàte for resists of either type. Htoo (37) showed that even after 30 min of air drying, however, there were still sufficient solvents present to drastically reduce the photosensitivity of negative resists (KTFR and KPR-2). Coatings air-dried overnight had the same photosensitivity as coatings with a postbake, which would indicate that oxygen absorption did not contribute significantly to the loss of photosensitivity after the 30-min dry.

Baking temperatures are important for the cyclized-rubber-base resists, particularly the newer resists, such as Waycoat IC, where the degree of cyclization has been controlled. This results in a resin that may flow slightly and soften at temperatures around 100°C or greater. Scumming may also occur. The earlier resists which usually had a higher degree of cyclization (such as KMER) are more resistant to elevated prebake temperatures and can usually be baked at temperatures of 100°C or slightly greater safely. Ester-base resist resins, such as KPR-2, can also be prebaked slightly above 100°C safely. Times of 15 to 30 min in natural convection ovens are adequate to expel solvents and do not appear harmful at the discussed temperature ranges. For the newer more temperature-sensitive resists, a temperature of 80°C for 20 min usually gives satisfactory results. Resists such as KMER or KTFR can be baked at 100 to 110°C for 15 min. Cyclized-rubber resists may be better baked in a nitrogen-gas atmosphere to reduce oxygen effects. The use of a nitrogen-purged oven during prebaking would provide the required ambient conditions.

Prebaking is also an essential step for positive resists, although air drying before baking does not appear as critical as for the negative resists. Baking at too high a temperature (>120°C) often results in loss of photosensitivity, although there is no reported oxygen effect as noted with the negative resists and an inert atmosphere appears to be of little advantage. In fact, it was indicated that nitrogen ambients during prebake were harmful because resist adhesion problems sometimes were noted (36).

Prebaking positive resists above 80°C is likely to cause some sensitizer decomposition to most resists. However, some decomposition is tolerable or possibly even desirable because the nature of the low-temperature thermal decomposition products is not known. Some ranges of time and temperature that may be useful in determining prebake times for positive resists are as follows:

Temperature (°C)	*Maximum time (min)*	*Suggested starting time (min)*
80	60	30
90	50	25
100	45	20
110	15	10
120	10	Rarely used

Obviously, some resists, such as AZ-111, may not tolerate temperatures above 100°C without scumming. Others, such as AZ-1350J or the Hunt positive resist, can be prebaked slightly above 100°C with good results. When the resist is to receive a high postbake, a high prebake temperature is often necessary for maximum adhesion. GAF suggests 70 to 80°C for 15 min for Microline resists.

Excessive baking cycles may promote the formation of pinholes that will show up in etching rather than lose a noticeable degree of photosensitivity. Some flowing of the coating due to changes in stress levels or other factors may also occur at higher temperatures, which could reduce the image quality slightly. Low-temperature prebaking results in retained solvents, and the symptoms of underexposure will be noted. Examples include poor image definition and incomplete developing in varying degrees.

A rapid method of prebaking either type of resist is the infrared (IR) oven (Fig. 7-5). Long-wave infrared radiant energy is particularly suited to drying photoresist films. There is considerable penetration through the resist film to the substrate, which becomes heated and tends to expel the retained solvents in the film from the inside out. The formation of a dried skin on the surface which retains solvents in the coating is thereby avoided, eliminating the need for air drying either type of resist. The conveyerized oven shown in Fig. 7-5 is capable of drying even thick resist coating in less than 1 min. One factor to be considered with this type of drying is the nature of the substrate surface beneath the resist coatings. Glossy metallic surfaces with low emissivity will tend to become hotter under the same conditions than dull, flat finishes, e.g., those on metallized unglazed thin-film substrates. Such factors are easily compensated for by varying the dwell time within or the temperature of the oven. It is mentioned as precaution to consider when drying substrates of mixed configuration with regard to surface finish, weight, and type of material.

The maximum temperature that the resist encounters in the IR oven, even for a short time, should be known or judged by the subsequent exposure and developing effects. A production test pattern or resolution pattern may be exposed and developed using standard techniques and times for the coating thickness and used to detect over- or underprebaking even if the actual temperature is not known. For example, the finest spaces opened up on a resolution pattern for the resist thickness can be used as the control, or standard of comparison. The dwell time in the heated zone (conveyer setting) and the power to the heaters can be varied until the finest spacing is obtained. It is important that all other process parameters not be varied during the test and that the coating thickness and uniformity be as close as possible from part to part. To ensure that the final image is acceptable, it is preferable to etch a pattern in the part. This will more readily detect fine scum, poor edge acuity, and related factors. To actually evaluate the settings, some prefer to simply try different settings until unsatisfactory results are obtained. If possible, the method of Htoo (37) can be used, in which the photosensitivity was actually measured. Times and temperatures were varied until maximum photosensitivity was obtained. An effective method that requires no unusual equipment is to construct a simple matrix in which the heater power, or some function thereof, is varied with the conveyer setting or dwell time in the oven. Test parts are processed over the entire range of both parameters and a range selected. Table 7-11 shows a sample matrix. As may be noted, several usable settings were obtained. In general, the more rapid conveyer settings should be used. In this case, a power setting of 4–6 with a conveyer setting of 10–15 would be satisfactory and would allow for slight process variations. Once the prebake is established, coating thickness can be varied to some extent to determine over what thickness range the setting is applicable. Usually, only slight variations are necessary to accommodate significant thickness changes.

Regardless of the method of prebaking, special consideration must be given to the prebake when the resist process does not employ postbakes. This is particularly true for positive resists. In addition to the prebake, adhesion is significantly improved when negative resists are exposed and thereby polymerized. However, positive resists undergo no such polymerization reactions, and the prebake becomes the sole adhesion-promoting step. In these instances, it is generally necesssary to prebake at the higher end of the temperature range. Most positive resists will stand a 90°C (194°F) prebake without losing photosensitivity, which will be adequate for many etching processes. Times of 15 to 30 min have been used for convection ovens, but actual cycles will depend on what the coating must endure.

EXPOSURE

Probably the single most important step in processing is exposure. After exposure, the resist coating is so different that the exposure step is often the reference point in the processing, e.g., preexposure baking or post-

TABLE 7-11 Example of a Matrix for Evaluating Infrared Conveyerized Method with Variable Power Settings

Power setting	Conveyor setting 5 (slowest)	10	15	20	25 (fastest)
2 (highest power)	Unsatisfactory, severe heat, fog	Unsatisfactory, severe scum	Unsatisfactory, scum	Marginal	Good
4	Marginal	Good	Good	Good	Marginal
6	Good	Good	Marginal	Unsatisfactory	Unsatisfactory
8	Marginal	Marginal	Unsatisfactory	Unsatisfactory, lost image	Unsatisfactory, lost image
10 (lowest power)	Marginal	Unsatisfactory, lost most of image	Unsatisfactory, lost image	Unsatisfactory, lost image	Unsatisfactory, lost image

RESOLUTION: 0.2 mil or less—good. 0.3 to 0.5 mil—marginal but acceptable. 0.6 mil and greater—unsatisfactory.

exposure baking. Negative resists depend upon exposure for adhesion and to minimize the deleterious effects of the developing cycle. Both positive and negative resists depend upon the exposure step to establish the image through photochemical reactions within the coating in the exposed areas. These reactions have been the subject of previous discussions in Chap. 2 and earlier in this chapter. For microelectronic applications, the primary need is for accuracy, and extra care is taken in this regard during exposure. In comparison to the previous applications where a line-width error of 0.2 or 0.3 mil was of little consequence (Chaps. 4 and 5), whole line widths of 0.3 mil are commonplace in microelectronics.

There are several methods of exposure used in microelectronic applications. Contact printing, in which the photomask image is in direct physical contact with the resist, is most commonly used and will be discussed herein. Off-contact printing, in which the mask image is spaced 20 μm or less from the resist surface, is rapidly increasing in popularity. Losses from the abrasive effects of particles of glass, silicon, or dust grinding between the mask and resist are reduced, and masks last four or five times longer. Projection printing utilizes highly advanced lens systems to project the image on the resist. Long mask life realized by this method would allow careful inspection and matching of masks. Although expensive, it may well become the predominant method in the future. Electron beam printers provide the ultimate in accuracy but are slow compared to the other methods and very expensive.

In Table 7-12, a generalized exposure process is presented based upon

TABLE 7-12 Contact Exposure Sequence

1. Check light-source intensity and distribution over printing surface (done periodically).
2. Prepare photomask for exposure. Clean to remove particulate matter (gaseous nitrogen, GN_2) and contamination such as resist spots and fingerprints, or replace mask if necessary.
3. Install photomask in printer. Ensure proper positioning and that the image or part number is correct.
4. Prepare resist-coated parts for exposure. Blow off surface with GN_2 if necessary, and inspect the condition of the resist coating. Scratched or damaged coating should not be exposed. Ensure that the part and the photomask image are of the correct part number in the proper layer sequence.
5. Align the photomask image to the part as required using nonactinic light sources. Ensure that proper contact is made between the part and the photomask at the completion of alignment.
6. Expose. Use the time range necessary for the thickness of resist and light intensity.
7. Upon expiration of exposure time, secure the actinic light; then eliminate any contact between the part and the photomask. Carefully remove the exposed part to avoid physical damage to the coating or mask. Store in the proper environment until development.

the writer's experiences. No specific piece of equipment is indicated. Devices as well as hybrid thin films are prepared by photoetching, and the comments are generally applicable in either instance for contact printing.

Exposure times for given positive- or negative-resist thicknesses are generally determined using resolution patterns, actual pattern line widths, or by step tablet. The last method is comparatively new but feasible (38), (68), (69). Resolution patterns (Fig. 7-6) containing line pairs of diminishing size are readily available commercially and very often used.

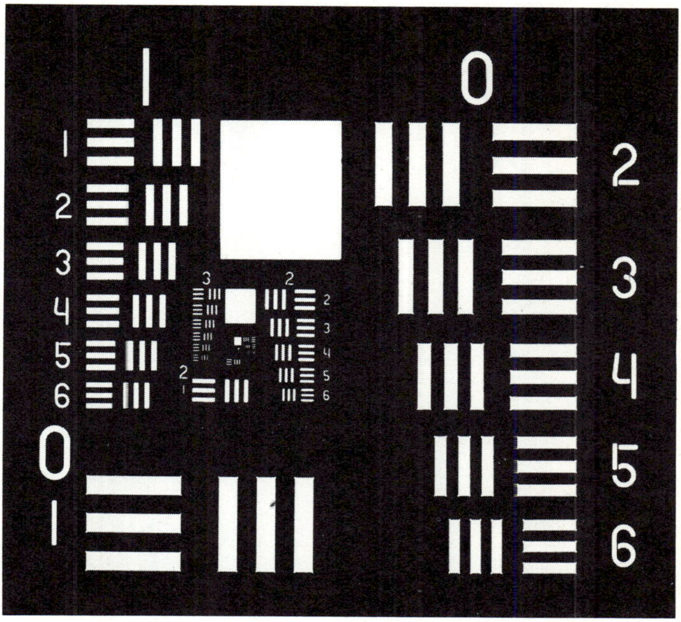

Fig. 7-6 Enlarged example of a resolution pattern based on the Mil Std 150A pattern. (*Avail. from Itek Corp., Lexington, Mass.*)

Exposure for maximum accuracy using these patterns is determined by trying various exposure times for a given resist thickness and light source while holding all other processing parameters as constant as possible. The exposure time that opens up the finest spaces in the pattern is generally optimum if it performs adequately during etching.

Accurate exposure times can also be determined by measuring a typical production pattern in lieu of the resolution pattern (Table 7-13). Again, the processing parameters and the resist thickness must be held as nearly constant as possible. Line widths and spacing deviations from the photomask are plotted as shown in Fig. 7-7 or in some related fashion and the optimum time determined. A problem associated with this method is

TABLE 7-13 Determination of Exposure Time by the Accuracy Method

1. Select resist thickness of adequate chemical resistance and accuracy (about 25% of minimum line width for starting point).
2. Expose for a series of times which encompass overexposure and the point of resist solubility.
3. Process through postexposure baking and examine images. Discard those with apparent processing flaws. Record deviation of resist image from photomask.
4. Etch parts using closely controlled etching conditions, and measure resulting deviations from both resist image and photomask.
5. Plot results and ranges as shown in Fig. 7-7, and select best exposure time.

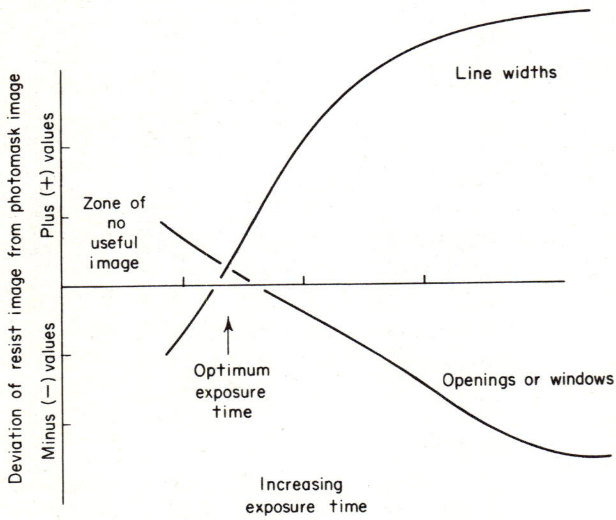

Fig. 7-7 Plot of pattern deviations versus exposure time.

ascertaining a "typical" pattern for the test. Perhaps as often as not, exposure time is determined simply by pattern inspection without measurement. A pattern is exposed for various times until the minimum spacings are obtained. The maximum allowable time which opens the spacings is then used.

Various optical effects work against obtaining accurate photoresist patterns. Probably the most noticeable effect is "diffraction," the spreading of light into areas behind an opaque object. Despite the concept that light travels in straight lines, it was shown long ago that a slight bending occurs around the edges of an obstruction, in this instance, the opaque areas on the photomask. In other words, light passing through a spacing will produce an illuminated area larger than that through which it passed.

Unfortunately, the light in the fringe or edge area of the pattern is nonuniform due to interference effects. If it were reasonably uniform, some form of compensation could probably be made. When two light waves of the same amplitude and wavelength arrive at a given point simultaneously and in phase, illumination will occur at that point. If they are out of phase, no illumination will occur because the waves annul each other. The width of the light and dark zones in the fringe varies directly as the wavelength of the light, and the contrast is most pronounced when the wavelengths are equal (monochromatic light). Light of this type produces total illumination in the light zones and none at all in the dark zones with an overall intensity decrease away from the original illuminated area. Ables and Geikas (39) showed that diffraction effects could be minimized by exposing with broad spectrum (nonmonochromatic) incoherent (out-of-phase) light sources with about 3° collimation. When reflected light from the silicon-SiO_2 interface of a wafer interacts with incident light within the resist layer under certain conditions, a phenomenon called "the standing wave effect" occurs (67). The result can be a series of steps or fringes in the resist images.

Low light intensity can also cause some unusual effects during resist exposure. Ideally, exposure time varies inversely as the light intensity. In other words, a more intense light source will require less exposure time for a given thickness of resist than one less powerful. This reciprocal relationship generally holds true for fairly thick layers of resist but fails on thin layers. Thin layers of some photoresists tend to require much longer exposure times for a given intensity level than would be predicted from thicker layers.

Reciprocity failure also has been noted with low-intensity light sources on resist thicknesses that exhibit reciprocity at moderate or high intensity. The reason for these failures is not clear, but with negative resists it most likely involves free-radical reactions with oxygen. Damon (40) presented data for KPR, KTFR, and KMER which indicated that some degree of reciprocity is experienced at an intensity of about 5 mW/cm^2 and greater. KPR shows considerably less sensitivity to decreasing light intensity than the other two resists. High light intensities may also have an effect in some instances. If the intensity reaches a very high level, there may be penetration of the finer lines on emulsion photomasks Altman (41) showed that the optical density of a 2-μm line is only about 1.0 under the conditions that an 8-μm line has a density of about 3.0. In other words, the 2-μm line transmits about 10 percent of the light, while the 8-μm line transmits only about 0.1 percent. If the density of the fine line is 0.5, which could happen because of variations in mask processing, about 32 percent of the light will be transmitted, which will cause significant processing problems. Small lines are lost for a variety of reasons anyway;

and when abnormal difficulties are encountered in determining the proper exposure, poor or variable photomask density should be considered as a possible source. Damon (40) indicated that minimum optical densities of 1.0 are adequate for light intensities of 10 mW/cm^2 or less. Very intense sources up to 25 mW/cm^2 could be accommodated by minimum densities of 2.0 or more.

A light spectrum can be varied through the use of optical filters to absorb undesirable ranges of radiation while transmitting the desired range. Infrared radiation is an example of undesirable radiation produced by the heated bulb surrounding an arc. One specific use of a filter cited by Damon is to remove that portion of the radiation spectrum to which emulsion photomasks were found to be most permeable (3000 to 3500 Å). Low transmission was noted after 3500 Å. However, such filters significantly reduced the light intensity.

Chromium masks did not suffer from the green-light leaks or the significant reduction in density with decreasing line widths. Table 7-14

TABLE 7-14 Some Advantages—Chromium Masks

1. Thinner than high-resolution photographic emulsions (0.1 μm vs. 4 to 8 μm), which allows finer lines without shadowing effect (diffraction and interference)
2. Shorter exposure times probable because of no emulsion in transparent areas
3. Greater optical density at the very edge of pattern lines
4. Exceptional chemical resistance and handling properties (see recleaning cycle)

lists some other advantages cited by Szupillo (42), and in Table 7-15 a recleaning cycle is given for resist-laden masks. Because of the expense, however, some consideration must be given in this regard to their procurement instead of emulsion masks. Secondly, chromium masks are more easily scratched by hard particles and epitaxial spikes than emulsion masks. The emulsion is thicker and softer and is able to absorb small particles that would damage hard chromium.

Collimation is an important consideration when one is printing accurate photoresist images. Light waves should impinge at approximately 90°

TABLE 7-15 Recleaning Procedure for Chromium Masks (42)

1. Immerse in hot photoresist stripper, swabbing as necessary.
2. Rinse with clean stripper.
3. Clean with detergent and water.
4. Rinse in distilled water using ultrasonic agitation (three changes of water).
5. Dry in water-miscible solvent vapor, e.g., isopropanol.

to the photoresist surface. If the waves impinge upon a mask badly askew, they will penetrate beneath the opaque areas on one side, and shadowing effects will occur on the opposite side. Correspondingly, poor image definition will result in the photoresist pattern.

Light radiates from a circular or point source in the form of radial lines in all directions. When a flat surface such as the photomask is placed in close proximity to the source, radials will strike the surface at very nearly a normal angle near the center with the angle of impingement (angle of incidence) increasing toward the edge of the mask. Obviously, the photoresist pattern in the middle will be an accurate reproduction of the mask image, but deviations will increase rapidly toward the edge of the pattern. However, if all but the few radials striking the center of the mask are shielded by an aperture and the part moved far enough from the light source, the entire mask will be exposed by the light at a near-normal angle. The result is an accurate photoresist image if the loss of intensity is not too great. Such collimation is called "distance collimation."

An obvious disadvantage is loss of intensity. The same light-flux density radiates through the aperture on a much larger area as distance increases from the source. In fact, the intensity decreases inversely as the square of the distance (an inverse-square law). As an example, the light intensity striking the mask at 36 in. is only about 11 percent of the intensity at 1 ft.

Collimation through the use of the lens system tends to reduce the rate of intensity loss with distance after the light has passed through the lens. The light radials striking the lens are bent by the lens to approximately parallel paths as they pass through. Hence, the spreading effect of the radials and, thereby, the associated intensity loss is reduced compared with simple distance collimation. While there is a significant loss of intensity involved in passing through the collimating lens, the intensity is normally greater than with distance collimation over the same focal area. Reflectors can greatly increase light output.

Schellenberg (43) indicated that the degree of undercutting of the opaque features on a photomask by a lens-collimated system depends upon the light-source size, the focal length of the lens system, and the quality of the lens system being used. It was indicated that dc mercury-vapor arcs actually have two hot spots which act as two small but separate sources which are projected by the lens system as overlapping circles. Light intensity varies within the area formed by the overlapping circles, but less than elsewhere. Information was presented to show that the part should be exposed within this area for best results. Elliptic reflectors were suggested as one method of improving the effectiveness of the mercury-vapor lamp by utilizing a much greater amount of the radiated energy.

Uniformity of intensity across the usable area is also a factor which can cause problems during exposure. Some lens collimating systems may cause overexposure in peripheral areas when adequate exposure is used for the center portion of a wafer or part. Distance-collimated light is generally quite uniform in this respect.

One disadvantage of very highly collimated systems is their tendency to reproduce any adventitious particulate matter on the mask in the photoresist pattern and the enhancement of refractive and interference effects. As stated earlier, Ables and Geikas (39) recommend approximately 3° collimation. Collimation is effectively reduced by poor and uneven mask-photoresist contact.

When using high-intensity light sources, photoresists may expose so rapidly that the difference between under- and overexposure may be a matter of only 1 or 2 s. To many, this represents no problem, but to some such rapid exposure is at least an annoyance. One method of overcoming the problem is the use of a neutral density filter, which reduces the light intensity that the resist will see and thereby increases the exposure time to a more acceptable level, where timing errors are not so critical. The filter may be placed over the mask frame holder in mask alignment systems or directly over the back of the photomask. They are available commercially with varying transmission levels on both glass and plastic base material.

After the exposure time has been determined, some method of maintaining the proper time is desirable. If the resist thickness is consistent, the energy required to expose it varies as the rate at which the energy is absorbed per unit area and the duration of the exposure. By taking the exposure time indicated by one of the aforementioned methods and measuring the rate at which the exposure source radiates light energy in the absorbable range, it is possible to obtain a constant relating to the quantity of energy required to repeat the exposure. In other words, the rate at which absorbable energy is being radiated per unit area P multiplied by the exposure time t is equal to a constant:

$$Pt = \text{constant}$$

It is not necessary to measure the actual rate at which energy is being radiated, only some function thereof. However, it is important to measure energy rates of that portion of the light-source spectrum to which the resist is known to be sensitive. For mercury-vapor light sources, measuring the line at 3660 Å will give a good indication for either positive or negative resists. Depending upon the sensitivity of the measuring instrument, it may be necessary to use a neutral-density filter to reduce the intensity to a level more accurately measured. Filters are necessary to isolate the portion of the spectrum around 3600 Å. For example, if the measured value for

P is 7.50 mW/cm^2 with a light source which had an optimum exposure time of 8 s the value of the constant is

$$\begin{aligned} K &= Pt \\ &= 7.50 \times 8 \\ &= 60.0 \text{ mW·s/cm}^2 \end{aligned}$$

This constant is useful to adjust the exposure time to account for slight variations encountered in day-to-day operations. The same type of resist must be used, and it must be applied under repeatable, controlled conditions. To control the exposure time, the value of P is measured, which is rapid and simple. This value is divided into the constant to get the adjusted exposure time. For example, if the measured value of P dropped to 7.0 mW/cm^2, the new exposure time is

$$\begin{aligned} Pt &= 60.0 \\ t &= \frac{60.0}{7.0} \\ &= 8.6 \text{ s} \end{aligned}$$

Using this method, no parts are used to adjust the exposure. Values for the constant are required for each type of resist and each distinct resist thickness. As the light source deteriorates with use, the value of the constant may change. Scheduled lamp changes may be monitored in this way. However, the constant should hold with at least a 10 to 15 percent variation in the value of P. The success of such a method is highly dependent upon the accuracy and repeatability of the measuring equipment and good measuring techniques.

Several factors are important in the selection of a light source and printing equipment to do a given job. In addition to financial reasons, technical considerations, such as light intensity and spectrum, the type and degree of collimation, and the intensity variation experienced across the usable surface are important. By far, the most accepted light source is the mercury-vapor lamp, which produces high-intensity radiation throughout the 3000 to 5000 Å range. Tungsten and quartz-iodide lamps and xenon or mercury-xenon lamps also produce some radiation within this range, but they are not as effective as high-pressure mercury-vapor sources. With the proper electrodes, carbon arcs are efficient producers of point-source radiation in the usable range. As the electrodes are burned and separated, however, there tends to be a variation in intensity and arc size. Probably this could be controlled to an acceptable level by an advanced form of servomechanism feeding the electrodes, but contamination from the electrode oxidation would also have to be controlled. All in all, the problems in using carbon arcs apparently outweigh any advantages over mercury-vapor lamps, and little use is made of them in microelectronics. Fluorescent and "black-light" units are fairly widely used in the circuit-board industry but not in microelectronics. Their broad physical expanse and comparatively low intensity undoubtedly cause collimation difficulties.

DEVELOPING AND RINSING

To develop microimages with negative resists, spraying is the preferred method. Spraying gives cleaner, more accurate images than immersion processes. The eroding effect of the spray helps remove partially exposed resist at the image edge and greatly reduces the tendency toward scum retention between lines or in windows. Spray developing is especially important for negative resists. Sprays should be highly atomized and at low pressure to avoid damage to the image. Sprayed developers are generally used only once, are always fresh, and can be filtered immediately prior to use. Considerable developer is lost, however, compared with immersion developing, and the fire hazard of vaporized solvent is substantially greater when using negative resists. One method that can be employed to reduce the hazard is to develop by immersion and use a solvent spray rinse. The developing solvents penetrate the resist to form a gel which is dissolved partially or eroded away by the spray-rinse action.

The actual negative-resist developing operation, as one may surmise, involves the selective dissolution of the unexposed areas. Chemically, the cross linking afforded by exposure does not alter the resist polymer enough to prevent a strong interaction with solvents capable of dissolving the unexposed un-cross-linked polymer. As a result, swelling of the exposed areas will occur. Swelling which can affect resist adhesion (44) if it is severe enough will continue until an equilibrium is reached. To minimize swelling, exposure must be adequate to ensure sufficient cross linking, which acts to restrain the swelling, and the developing solvent must not be too active. In fact, the poorer the developing solvent, the lower the swelling but the greater the probability of scum formation from underdeveloping. It is usually necessary to compromise and use a solvent or solvent mixture that has a moderate amount of solvency. Most negative photoresists have developers prepared by the manufacturer which are prefiltered and suitable for microwork. Mixtures of xylene and Stoddard solvent are often employed for developing the cyclized polyisoprene-base resists. Thin coatings can be developed with a high percentage of the latter, which is the poorer solvent of the two. Martinson (8) indicated that 1 μm of KTFR coatings could be developed with Stoddard solvent alone. From patent literature, it is seen that developers for the cinnamate resists contain hydrocarbon solvents and an active solvent. Xylene (four parts) and cyclohexanone (one part), xylene and Cellosolve acetate, and others (45) are examples. Taylor (26,27) indicated that dipropylcarbonate mixed with the developer in a 40 : 60 ratio gives good results with KMER. Following the developing of negative resists, a rinse is used to remove the excess developing solvents and the undissolved resist from the part. Most rinses following the developing of negative resists are now nonaqueous. After developing, the negative-acting resist image is swollen with absorbed solvents and is susceptible to damage or lifting around the edge during rinsing. Water has a very strong polar attraction for the silica on the wafer surface and may tend to undermine the swollen image and promote adhesion loss. Normal butyl acetate is widely favored as a rinse immediately following developing. It is applied both by spraying and immersion with agitation to mechanically remove the unexposed areas, has little affinity for the silica surface, and is a relatively poor solvent for the resist resin while having some solvency for the absorbed developer. Alcohol (methanols, ethanol) may be mixed with the butyl acetate, supposedly to give better edge acuity.

Mixtures of the lower aliphatic alcohol and developer are sometimes used as rinses for both the ester- and hydrocarbon-resist negative-resist resins after developing. Improved image definition is claimed. Straight alcohol rinses after developing are not recommended by Kodak (8) because of possible damage to the image by severe surface shrinkage and reticulation (wrinkling).

Solvent spraying with the proper developer for 20 to 30 s will give acceptable results for most applications. Rinses prepared by mixing 60 to 80 vol percent developer with ethanol can be used on both cinnamate- and isoprenoid-resist resins. Kodak Micro resist rinse is available and gives good results. Hunt also has a similar product. Spraying for 20 s is usually adequate following developing.

Thinners are sometimes employed in the developing cycle of the polyisoprene resists. Generally this should be considered carefully. Thinners contain a higher percentage of aromatic hydrocarbons than the developers and can cause greater swelling of the image and possible damage.

Water in the developing solvents may cause scumming with negative resists, particularly on silica surfaces. Water tends to concentrate at the silica surface of wafers during developing where the local concentration of resin in the developing solvent is high, and, as a nonsolvent, it can cause precipitation in the form of a thin resin scum under some conditions. Commercial-grade solvents which could be used in the developing process, e.g., xylene and Stoddard solvents, tend to have considerable variation in water content. Relatively dry solvents may also acquire ambient moisture after some time in open air. Chlorinated hydrocarbon solvents have a much lower water content generally but are difficult to employ in a developing solvent.

The situation with positive resists during developing is different in a number of ways, particularly with regard to the effect of water. There is little or no swelling of the image area by the aqueous developer, although prolonged immersion may cause physical damage. Water has less tendency to precipitate the nonimage areas and cause scum formation and is actually a solvent for the developed residues which contain acid salts. Adhesion loss is minimized by the lack of swelling and reduced solvent action such as moisture penetration through the swollen image to the silica surface.

As indicated earlier physical damage can result from overdeveloping, perhaps more so than with negative resists. The active ingredient in the developing solutions is alkaline in nature, and developing cycles for some resists (Shipley) can be controlled by measuring the normality of the developer and adjusting the developing time accordingly. Thin coatings of resist may, in fact, process better with less pinholing at a more dilute range. It should be noted that with use the normality of the developer decreases through neutralization of acid groups formed during exposure of the photosensitive elements (see Fig. 2-18). Stockley (36) indicated that the standard AZ developer solution should be discarded when the alkaline normality falls below 0.205 from its original 0.24 normal concentration. It was further indicated that AZ-303 developing solutions should not be used below an alkaline normality of 0.29 *N*. The original makeup varies between 0.329 and 0.339 *N*. When very thin resist coatings are used, it may be desirable to adjust the normality of the solutions to a lower alkalinity level to obtain the best results. For example, very thin coatings of AZ-111 could be developed at a normality of 0.20 or lower. Concentrated developer solutions may produce good results with shortened exposures.

For most applications, adequate results can be obtained by immersion developing with agitation. Spray developing is also quite useful and preferred by many users. Most positive resists develop cleanly in 20 to 40 s by either method. AZ-1350B coatings normally require only about 5 s.

After developing and before water rinsing, a rinse prepared by greatly diluting the developing concentrates (10 to 20 parts water for each part concentrate) may be used to help clean up the image if desired, but this is usually unnecessary. Thorough water rinsing is essential in either instance.

POSTBAKING

Because exposure methods favor accuracy and thereby minimum times and because of the nature of the surfaces to which the resist must adhere, postbaking is an essential step in the processing of negative resists for microelectronic applications. Exposure still leaves extensive chemical unsaturation available for thermal polymerization in negative resists. Both positive and negative resists gain (1) greater overall chemical and pinhole resistance and (2) improved surface adhesion by the utilization of adequate postbaking. Postbaking also expels retained volatiles from the developing cycle in both resists and provides energy for realignment of the resist molecules to attractive sites on the substrate surface (46). The resins in some positive resists may also be capable of forming three-dimensional polymers upon heating, although there is a lack of general information on thermally induced chemical reactions between the components of positive resists.

TABLE 7-16 Postbaking Cycles

Resist	Use	Postbake cycle	Reference
Waycoat IC	General etching	135°C for 15 min in GN_2	Hunt Tech. Bulletin 35
Waycoat SC	Deep etching	150–175°C for 25 min	Hunt Tech. Bulletin 23
KMER	Silicon etching	160–200°C for 15 min	(7)
KTFR	Chromium mask making	100°C "short time"	(65)
KMER	Oxide etching	120°C for 10 min followed by 180°C for 30 min	(66)
KMER	Resist evaluation	140°C	
KPR	Resist evaluation	190°C	
KPR-2	Resist evaluation	190°C	(5)
KTFR	Resist evaluation	140°C	
Four parts KPR, one part KPL	Resist evaluation	190°C	
KTFR	Silicon etching	160°C for 10–20 min	(8)
Micro resist 747	General etching	130°C for 10–30 min	Kodak Data Sheet p. 306
AZ-1350B	General etching	110°C for 30 min	Author
AZ-1350J	General etching	150°C for 30 min	
AZ-111	General etching	110°C for 30 min	
GAF Microline PR-102	General etching	80°C for 30 min	
Waycoat Positive HR	General etching	100°C for 10 min	Hunt Tech. Bulletin 34

Excessive postbake cycles are characterized by one or more of the following: (1) image distortion, (2) loss of adhesion, (3) brittleness, (4) decomposition, and (5) stripping difficulties. From a practical standpoint, the postbake should be the minimum required to reliably obtain the necessary adhesion and chemical resistance for subsequent processing. Otherwise, excessive effort may be required to strip the resist, which would bring about unnecessary costs. Some typical postbake cycles are shown in Table 7-16. In addition to the cycles shown, the Shipley resists can be baked up to 200°C for 30 min when extreme chemical resistance is required. Stripping is difficult, however.

STRIPPING

Stripping is an often overlooked but significant contributor to yield and cost of component manufacture. Incomplete removal of resists most assuredly causes problems in subsequent photoresist operations (47) and thermal processing of wafers (48). Because of the delicate nature of the processed parts, nearly any form of rubbing or scrubbing has to be avoided; this complicates the resist removal process. Aluminuim, the common conductor material in microelectronics, is easily corroded in thin layers; this must also be considered when stripping methods are selected. However, even the slightest of resist films remaining on aluminum or other metal surfaces can have a deleterious effect on subsequent wire bonding operations, making complete removal very important. Sodium ion concentration is also important for some strippers when one makes devices sensitive to ion-migration phenomena. Examples of resist stripping solutions are shown in Table 7-17.

Resist coatings can be removed using oxygen plasmas which oxidize the resist to volatile compounds, e.g., water and carbon oxides. Oxygen-containing gases at low pressure are passed through an RF field to produce atomic oxygen, electrons, and various other molecules and atoms, some of which are in the excited state. The reactive substances in the plasma convert the photoresist film into the aforementioned volatiles, which are removed from the chamber. Such devices are available in various sizes, some of which are shown in Figs. 7-8 and 7-9.

Upon volatilization of the organics, inorganic impurities are left on the surface; they must be removed by some means. Irving (49) indicated that the use of dilute HF was excellent for this purpose. One advantage of the dry method of resist removal is the lack of dependence on the type of resist or prior processing. Even the toughest most thoroughly cured resist films are removed with comparative ease at a rate of 1,000 to 2,000 Å/min. Temperatures of the parts usually remain below 150°C, and most metals are not significantly affected by the plasma bombardment. However,

TABLE 7-17 Resist Stripping Solutions

Resist	Removal method	Reference
All	Soak in hot (80–100°C) commercial strippers such as Hunt Microstrip	65,24,36, 70
KMER, KTFR	Triethylenetetramine (hot)	65,32
KPR	Equal volumes sulfuric acid and hydrogen peroxide (30%)	65,37
Negative resists in general	Trichloroethylene presoak plus commerical stripper	66,35
Negative resists in general	Hot, concentrated sulfuric acid	None
Negative resists in general	Concentrated sulfuric acid saturated with chromic acid	None
Positive resists in general	Acetone DMF (hot) MEK MIBK Dioxane Butyl Cellosolve (hot) or Butyl Carbitol Aqueous caustics after image exposure (UV light)	

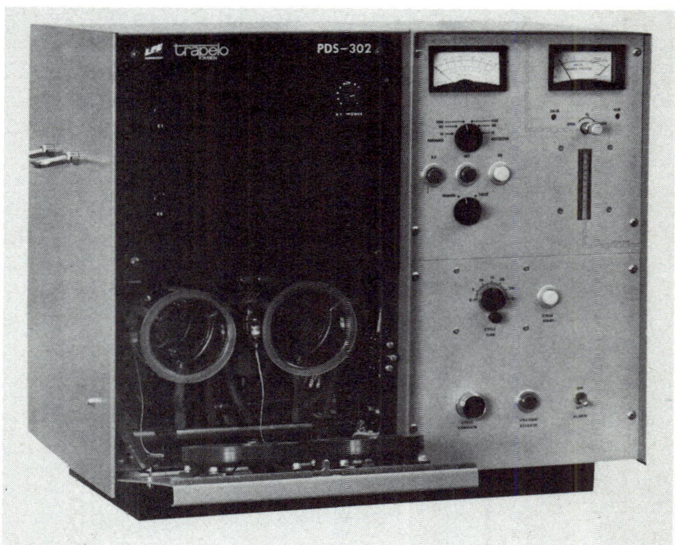

Fig. 7-8 Low-temperature plasma unit for dry-photoresist removal. (*LFE Corp., Trapelo Div., Richmond, Calif.*)

Fig. 7-9 Dry-photoresist removal equipment with a large chamber. (*LFE Corp., Trapelo Div., Richmond, Calif.*)

some exceptions are copper and Nichrome resistors. The resistance values of the latter are often affected, presumably through the conversion of the resistor metal to oxides, and copper will oxidize.

PINHOLE TESTING AND PROCESSING EVALUATION

One method used to compare process variables and/or photoresists is the so-called "pinhole tests." These tests involve the formation of a stable, uniform SiO_2 layer on silicon wafers, which would normally protect the silicon from attack by selective etchants. To test the effectiveness of the photoresist, or processing variables with a given resist, a coating is applied and processed followed by a timed immersion in an etchant for SiO_2. If weak spots or defects are present in the photoresist layer, the unprotected SiO_2 will be etched through to the silicon, provided the immersion time is sufficient. Thereafter, etchants that attack silicon but not SiO_2 will enter through the pinholes and etch enlarged, visible areas, which can be visually counted. Figure 7-10 shows a test sequence for using the pinhole tests to compare photoresists.

Pinhole testing alone is of some value but useless if the resist thickness is such that pattern resolution is poor. It is therefore common practice to

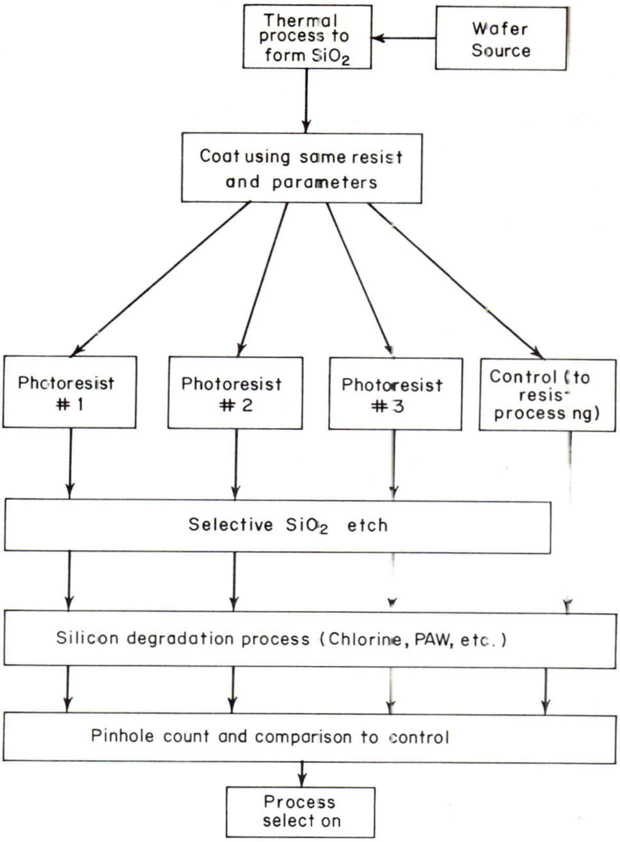

Fig. 7-10 Use of a pinhole test to compare resists or process variables.

use the test in conjuction with the resolution capabilities of the resist at the thickness applied. Adhesion as related to undercut can also be indicated as well as scum formation tendencies if a resolution pattern is etched in the SiO_2 along with the pinhole testing. Results must be considered collectively to be meaningful.

Generally, the SiO_2 etchants are ammonium acid fluorides. As an example:

Seven volumes electronic-grade ammonium fluoride solution (40 wt percent)
One volume electronic-grade hydrofluoric acid (49 wt percent)

Parts are immersed longer than necessary to remove unprotected SiO_2, slightly in excess of the time to etch a fine-line pattern in the oxide thickness. The resist is applied, dried, exposed, developed, and postbaked according to a set schedule. It is important that all steps that are necessary for image formation be followed except, of course, that no image is formed. Care should be taken to avoid contamination during exposure; this would produce pinholes in the resist. Once the SiO_2 etch is completed, the parts are rinsed thoroughly in deionized water and dried. Thereafter, the photoresist coating is stripped, and the wafers are cleaned and dried. If the back of the wafers is not protected during the SiO_2 etch it is advisable to apply a protective lacquer to prevent damage from immersion processes. For thermal processes, the SiO_2 offers the primary protection, and the lacquer must be applied before etching, preferably before postbake if screening inks are used.

Methods of etching the silicon through the pinholes in the SiO_2 are fairly numerous. They can employ (1) nitric-hydrofluoric acid, (2) hot chlorine gas, (3) pyrocatechol-amine-water, (4) caustic-water, (5) electrographic techniques, (6) electrical capacitance breakdown voltage, and (7) microscopy. Each has its advantages. Nitric-hydrofluoric acid is a common available mixture used to etch silicon. Mixtures such as 95 : 5 pbv nitric and hydrofluoric acids, respectively, are useful. Immersion times of 15 min at 23°C are generally adequate to enlarge the small pinholes exposing the silicon to a readily visible size. Thorough rinsing should, of course, be used. This method is rapid, uses readily available chemicals, and is a simple variation in daily routine. Masking the back of the wafer may or may not be necessary, depending upon the thickness of the silicon and whether the SiO_2 back was used for an etching time indicator during the pinhole-forming etch.

Hot-chlorine-gas etching (50,51,5) is a recognized method of pinhole testing. Basically, the parts are subjected to a flow of hot chlorine gas ($\approx$900°C) in a tube furnace for a short time ($\approx$2 min). Chlorine selectively etches the silicon through the SiO_2 and enlarges the defect to readily detectable size. Table 7-18 shows briefly the procedure from the testing by Kelly (5).

Sullivan (52) indicated that the pyrocatechol-amine-water silicon etchant gives higher pinhole resolution than the chlorine-gas method although it is much slower. An etching time of 6 h was reportedly typical at 115°C,

TABLE 7-18 Chlorine-Gas Silicon Etch for Pinhole Detection (5)

1. Heat tube furnace to 910 $\pm$ 10°C, and flush nitrogen gas for 5 min at 1.0 ft^3/h.
2. Continuing gas flow, insert the wafers contained in a quartz boat, and continue heating for 4 min or until wafers reach 910°C.
3. Stop nitrogen-gas flow and start flowing chlorine gas at 0.6 ft^3/h. Continue chlorine flow for 2.0 min and stop. Start flushing with nitrogen.
4. Flush with nitrogen for 5 min, remove wafers, cool, and count pinholes.

which is near boiling for the solution shown in Table 7-19. This method was investigated earlier by Crishal and Harrington (53) and by Finne and Klein (54).

Silicon is etchable in caustic solutions, such as potassium or sodium hydroxide. As an example, 25 percent solutions of sodium hydroxide at 80 to 100°C will usually give adequate silicon etching of pinholes in 10 to 20 min.

TABLE 7-19 Pyrocatechol-Amine Water Silicon Etchant for Pinhole Detection

Compound	Chemical formula	Amount per liter
Pyrocatechol (1,2-benzenediol or 1,2-dihydroxy-benzene)	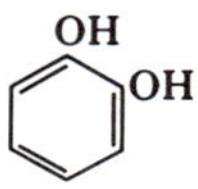	120 g
Ethylenediamine (1,2-diaminoethane)	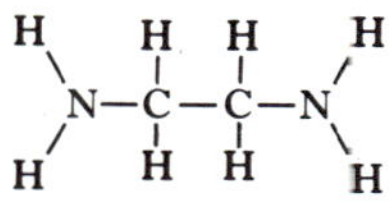	680 mils
Water	....	320 mils

Electrographic methods involve the application of an electric current to promote the formation of visually indicating phenomena from or in the vicinity of the pinhole. One such method developed by Lee, McCloskey, and Licari (55) involves the anodic oxidation of a colorless reagent contained in a porous material, such as filter paper, when it is in contact with the SiO_2 surface. Pinholes permit sufficient current flow to oxidize the reagent (e.g., benzidine) to a colored form on the filter paper. The method is fairly rapid and nondestructive; it can be used on completed parts in some instances and affords a permanent record of the test. Another method of this type developed by Besser and Meinhard (56) detects pinholes through the evaluation of a continuous stream of bubbles from pinholes as well as the deposition of a copper halo around the defect. The method setup is shown in Fig. 7-11. A dc potential of 25 to 50 V typically is applied, which results in a stream of hydrogen gas from pinholes in the SiO_2, although current flow is in the milliampere range. One advantage of this method is the magnification of the defect by the copper halo, which allows easier counting.

In all the aforementioned methods, counting methods are required for repeatable results. Magnification, the accuracy of the count, the number of holes that the method will actually highlight, and other factors may lead to

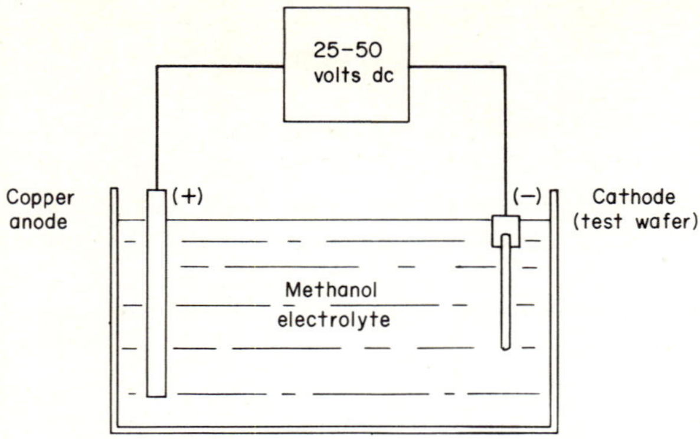

Fig. 7-11 Pinhole detection by gas evolution (56).

erroneous conclusions regarding the effectiveness of the photoresist in preventing pinholes. Hagenbuck (57) indicated the following points as possible sources of counting errors:

1. Pinholes that occur in a straight line should not be counted. These were probably caused by scratches in the SiO_2 surface.
2. Pinhole clusters should not be counted. These could have been caused by contamination during processing or damage to either the resist or SiO_2 film.

Kelly (5) indicated that pinholes within 2 mm of the wafer edge should not be counted either. Areas measured vary from the whole surface of a wafer to 0.25 cm^2, probably necessitated by the resolution of the test method.

Another method for detecting pinholes and defects is based on the change in capacitance brought about by defects in the dielectric SiO_2 layer (58). The bottom plate of the capacitor (Fig. 7-12) is formed by the silicon layer

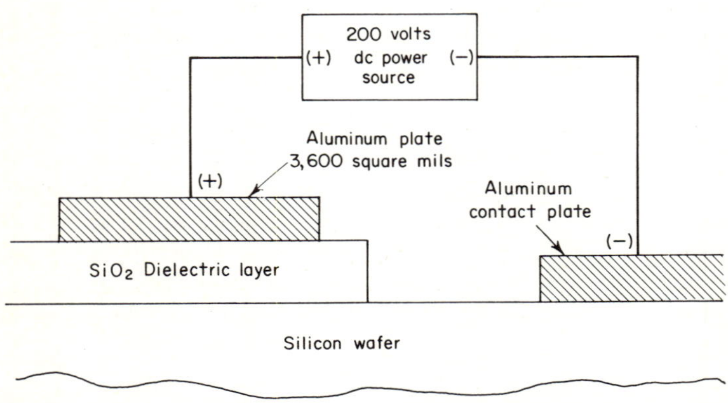

Fig. 7-12 Example of a capacitance method of SiO_2 pinhole testing (58).

and is picked up by aluminum electrodes deposited on the silicon through windows etched in the SiO_2. Defects in the photoresist film form defects in the dielectric SiO_2 during the window etching. The capacitor is formed when the top plate is deposited. Either deposition through a mask or photoetching can be used to form the aluminum pattern. In the latter instance etchants that do not affect the SiO_2 layer must be used. When the capacitors are formed and electrically tested against a criterion, the percent yield of the capacitors reflects the damage to the SiO_2 layer. This method does not require the careful counting techniques of the other methods and accounts for pinholes not completely etched through. It is useful as a check of other methods because of its accuracy. Investigations (59,60) regarding the effect of voltage polarity on the dielectric breakdown of thin oxide films on silicon have been made and may be of interest to those pursuing this method of pinhole detection.

Direct detection of defects by observation at high magnification is a useful nondestructive method of detecting defects, especially when dark field lighting is used. The method can be used routinely as a process-control tool, or electron microscopy can be used to isolate and examine specific defects. It is, of course, an integral part of most of the other methods, which simply eases the chore of defect counting by increasing the defect size for ease of detection.

Some other methods tried with varying degrees of success include thermal measurements, photographic techniques, low-pressure gas discharge, and biological types of dye staining and dye penetrants.

Thickness measurements for resist films have not been emphasized in this chapter but may be of interest in pinhole studies and other process evaluations. The performance during subsequent processing and the image resolution as related to repeatable application parameters have been emphasized heretofore.

Resist-film thickness can be determined in a number of ways. One simple method is to roughly estimate it from the weight change of the part due to the applied resist film. The average thickness of a coating is

$$t_{avg} = \frac{\text{weight of resist coating}}{\text{coated area} \times \text{density of coating}}$$

$$= \frac{W}{Ap}$$

When the weight is given in grams, the area in square centimeters, and the density in grams per cubic centimeter, the average film thickness in micrometers is

$$t_{avg} = 10^4 \frac{W}{A} \qquad \mu\text{m}$$

A density of 1 g/cm^3 was assumed for the dried film. The formula simplifies further if parts with approximately the same area are used. For best results, care must be taken to ensure that the weight has stabilized.

A common method utilizes a stylus recording instrument. P. A. Hunt research used the Sloan Dektak stylus recording instrument for determination of resist-film thicknesses (61). Sodium-line interferometers have been used with a vapor-deposited aluminum film over the step representing the film thickness. Reflectance measurements (62,63) are also used. Warnecke (64) described a promising method using an interferometer.

REFERENCES

1. R. K. Agnihotri, D. L. Falcon, F. P. Hood, L. G. Lesoine, C. D. Needham, and J. A. Offenbach, *Tech. Pap., Reg. Tech. Conf. Soc. Plast. Eng., Mid-Hudson Sec.*, October, 1970, p. 77.
2. Symposium, P. A. Hunt Company, no published proceedings, Palo Alto, Calif., Jan. 27, 1971.
3. H. L. Fisher, U.S. Patent 1,605,180 (1926).
4. M. Hepher and H. M. Wagner, U.S. Patent 2,852,379 (1958).
5. R. F. Kelly, *Proc. Kodak Semin. Microminiaturization, Kodak Publ. P*-77, 1965, p. 38.
6. K. G. Clark, *Solid State Technol.*, vol. 14, no. 6, June, 1971, p. 52.
7. L. E. Martinson, *Proc. Second Kodak Semin. Microminiaturization, Kodak Publ. P*-89, 1966, p. 31.
8. L. E. Martinson, *Proc. Kodak Semin. Microminiaturization, Kodak Publ. P*-77, 1965, p. 49.
9. W. Kornfeld, *Proc. Second Kodak Semin. Microminiaturization, Kodak Publ. P*-89, 1966, p. 23.
10. H. A. Levine, L. G. Lesoine, and J. A. Offenbach, *Kodak Photoresist Semin. Proc., Kodak Publ. P*-192-*A*, vol. I, p. 15, 1968.
11. F. J. Fauner and L. E. Martinson, U.S. Patent 3,493,380 (1968).
12. E. M. Robertson and W. West, U.S. Patent 2,732,301 (1956).
13. J. Tulumello and W. B. Harding, *Plating*, vol. 54, no. 11, p. 1234, 1967.
14. J. D. Guttenplan, D. Maslow, and W. S. DeForest, *Proc. Fall Hybrid Microelectronics Symp.*, Boston, Oct. 31, 1967.
15. J. V. Brandewie and C. W. Scott, *NASA Rep. No. NASA-CR*-117904, 1970.
16. C. K. Monzeglio, *Solid State Technol.*, vol. 15, no. 2, 1972.
17. *ASTM Spec. Tech. Publ. No.* 246, 1958.
18. R. W. Lee, *Physics and Chemistry of Glass*, vol. 5, no. 35, 1964.
19. J. E. Meinhard, *Electrochem. Soc. Spring Meeting, Cleveland, Recent New Pap. No.* 19, 1966.
20. T. E. Burgess and F. M. Fawkes, *Extended Abstracts Electronic Div. Electrochem. Div. Electrochem. Soc.*, vol. 15, no. 1, p. 110, 1966.
21. G. F. Damon, *Proc. Second Kodak Semin. Microminiaturization, Kodak Publ. P*-89, 1966, p. 36.
22. G. C. Schwartz, *Proc. Kodak Semin. Microminiaturization, Kodak Publ. P*-77, 1965, p. 41.
23. K. G. Clark, *Solid State Technol.*, vol. 14, no. 9, September, 1971, p. 48.
24. S. A. Harrell and G. M. Streater, *Kodak Photoresist Semin. Proc., Kodak Publ. P*-192-*A*, vol. I, p. 20, 1968.

25. S. A. Harrell, *Proc. Second Kodak Semin. Microminiaturization, Kodak Publ., P*-89, 1966, p. 50.
26. C. J. Taylor, U.S. Patent 3,458,313 (1969).
27. C. J. Taylor, *Proc. Kodak Semin. Microminiaturization, Kodak Publ. P*-77, 1965, p. 17.
28. W. S. Bryant, *ibid.*, p. 33.
29. R. Steinman, U.S. Patent 2,688,006 (1954).
30. R. Steinman, U.S. Patent 2,688,007 (1954).
31. T. A. Te Grotenhuis, U.S. Patent 2,742,378 (1956).
32. T. A. Te Grotenhuis, U.S. Patent 2,841,566 (1958).
33. T. E. White, *20th Annual SPI Preprints*, February, 1965.
34. R. G. Bartfeld, *Kodak Photoresist Semin. Proc., Kodak Publ. P*-192-*B*, vol. II, p. 30, 1968.
35. R. H. Collins and F. T. Deverse, U.S. Patent 3,549,368 (1970).
36. F. Stockley, *Proc. EIPC Conf. Printed Circuit Manufacturing*, Hargen, Switzerland, February, 1970, p. 93.
37. M. S. Htoo, *Kodak Photoresist Semin. Proc., P*-192-*A*, vol. I, p. 25, 1968.
38. I. F. Stacy, *Proc. Kodak Photoresist Semin., Kodak Publ. P*-209, 1969, p. 64.
39. G. I. Geikas and B. D. Ables, *Kodak Photoresist Semin. Proc., Kodak Publ. P*-192-*B*, vol. II, p. 47, 1968.
40. G. Damon, *ibid.*, p. 26.
41. J. H. Altman, *Solid State Technol.*, vol. 12, no. 7, July, 1969, p. 34.
42. R. E. Szupillo, *Solid State Technol.*, vol. 12, no. 7, July, 1969, pp. 49–52.
43. H. Schellenberg, *Proc. Kodak Photoresist Semin., Kodak Publ. P*-209, 1969, p. 54.
44. E. B. Davidson, *Tech. Pap., Reg. Tech Conf. Soc. Plast. Eng., Mid-Hudson Sec.*, October, 1970, p. 141.
45. L. M. Minsk, W. P. van Deusen, and E. M. Robertson, U.S. Patent 2,670,286 (1954).
46. D. H. Gittleman, *Circuits Manufacturing*, October, 1970, p. 52.
47. K. R. Dunham, *Solid State Technol.* vol. 14, no. 6, June, 1971, p. 41.
48. M. L. Long and S. A. Harrell, *Proc. Kodak Photoresist Semin., Kodak Publ. P*-209, 1969, p. 33.
49. S. M. Irving, *Kodak Photoresist Semin. Proc., Kodak Publ. P*-192-*B*, vol. II, p. 26, 1968.
50. S. W. Ing, R. E. Morrison, and J. E. Sanders, *J. Electrochem. Soc.*, vol. 109, p. 221, 1962.
51. E. F. Duffek, E. A. Benjamini, and C. Mylroie, *Electrochem. Tech.*, vol. 3, p. 75, 1965.
52. M. V. Sullivan, *Proc. Kodak Semin. Microminiaturization, Kodak Publ. P*-77, 1965, p. 30.
53. J. M. Crishal and A. L. Harrington, *J. Electrochem. Soc.*, vol. 109, p. 71C, 1962.
54. R. M. Finne and D. L. Klein, *J. Electrochem. Soc.*, vol. 111, p. 63C, 1964.
55. S. M. Lee, J. P. McCloskey, and J. J. Licari, *Insulation*, vol. 15, p. 40, 1968.
56. P. J. Besser and J. E. Meinhard, *Proc. Symp. Manufacturing In-Process Control and Measuring Techniques for Semiconductors*, Phoenix, Ariz., vol. II, p. 16-1, 1966.
57. E. G. Hagenbuck, *Rep. PR*-161-1, P. A. Hunt Company, 1971.
58. E. P. Donovan, *Proc. Kodak Semin. Microminiaturization, Kodak Publ. P*-77, 1965, p. 27.
59. F. L. Worthing, *J. Electrochem. Soc.*, vol. 115, no. 1, 1968.
60. N. Klein, *IEEE Transactions on Electron Devices*, vol. ED-13, no. 11, 1966.
61. E. G. Hagenbuck, *Research Rep. PR-Z*-11*a*, P. A. Hunt, Chemical Company, December, 1971.

62. E. A. Corl and H. Wimpfheimer, *Solid State Electronics*, vol. 7, p. 755, 1964.
63. L. A. Murry, N. Goldsmith, and E. L. Jordan, *Electrochem. Tech.*, vol. 4, p. 508, 1966.
64. A. J. Warnecke, *Proc. Kodak Photoresist Semin., Kodak Publ. P*-209, 1969, p. 83.
65. R. D. Barnes, *Proc. Kodak Semin. Microminiaturization, Kodak Publ. P*-77, 1965, p. 6.
66. J. C. Lawson, *ibid.*, p. 22.
67. D. F. Ilten and K. V. Patel, *Image Technol.*, February/March, 1971, pp. 9–14.
68. R. K. Curran and R. E. Kerwin, U.S. Patent 3,631,772 (1972).
69. M. R. Goldrick and L. R. Plauger, *Photogr. Sci. Engr.*, vol. 17, no. 4, July/August, 1973, p. 386.
70. J. J. Berilla, J. H. Birk, et al., U.S. Patent 3,582,401 (1967).

Index

Index